Contents

Preface   4

SKILLS PRACTICE

1. The Calculator   6
2. Rounding   8
3. Paying for Adverts   10
4. Food: The Calorie Counters   11
5. Fractions (one)   13
6. Fractions (two)   15
7. Fractions (three)   18
8. Decimals   21
9. Changing Fractions to Decimals   25
10. Percentages   27
11. Weights and Measures   30
12. 12 and 24 hour clock times   34
13. Ratio and Proportion   37
14. Graphs and Charts   39
15. Areas and Perimeters   42
16. Volumes   45
17. Scales and Maps   48
18. The City Hall   50
19. The Garden   51
20. The Rag Trade   52
21. The Bakery   53
22. Angles, Gradients and Bearings   55

MONEY

23. Wages and Salaries   59
24. Bank Accounts   62
25. Savings   63
26. Hire Purchase   65
27. VAT and Service Charges   68
28. Balance Sheets   71
29. The Holidays   73
30. Holiday Exchange Rates   75

HOUSEHOLD

31. Buying a House   78
32. Household Insurance   82
33. The Rates   83
34. The Electricity Bill   85
35. The Gas Bill   89
36. Telephone Charges   92
37. D.I.Y.   94

CAR

38. Buying Cars   97
39. Depreciation and Running Costs   99
40. Car Insurance   101
41. Car and Van Hire   104

HUNTINGDON
TECHNICAL COLLEGE
LIBRARY

Preface

Everyday Maths is designed as a consolidation book for the older CSE pupil and the 16–19 year old student following City and Guilds Numeracy or RSA Arithmetic in Further Education Colleges. It should also be of interest to trainees on Y.T.S. Schemes in Further Education Colleges and other centres.

I believe I have provided sufficient revision in the first chapter. This is developed by using everyday applications of these skills in the remainder of the book. Where possible I have made suggestions for discussion and project work in order that the material should be suitable for Schools and Colleges where life skills courses are under development.

The calculator used throughout was a Casio which has the facility to find percentages, mark up or give discount by pressing the %, + or − keys. For certain other makes the = key needs to be pressed also. To avoid confusion I have shown the = key in my worked examples as this makes no difference to the result obtained on the Casio.

A.L.

Acknowledgements

The author and publishers would like to thank the following individuals and organisations for permission to reproduce photographs in this book:

Casio Electronics Co Ltd (Page 6); Berger Decorative Paints (Page 9); Cambridge Evening News (Pages 8, 17, 27); Billing Boats A/S, Denmark (Page 15); British Broadcasting Corporation (Page 17); George Williams (Pages 19, 69, 75); Sony (UK) Ltd (Page 28); Alitalia (Page 34); Eastern Counties Omnibus Co Ltd (Page 36); Bar O Mix Ltd (Page 37); St Albans District Council (Page 50); British Rail Engineering Ltd (Page 59); Halifax Building Society (Page 63); Bang and Olufsen UK Ltd (Pages 65 and 67); ABI Design and Marketing Ltd (Pages 67 and 76); Exide Batteries (Page 68); Custom Ltd (Page 69); Thomson Holidays (Page 73); J B Briggs (Page 75); Renault UK Ltd (Pages 76 and 98); Hertfordshire County Fire Brigade (Page 82); Central Electricity Generating Board (Page 85); Osram (GEC) Ltd (Page 87); Hotpoint Limited (Page 87); Philips Electronic & Associated Industries Ltd (Page 87); TI Creda Limited (Page 88); Lec Refrigeration PLC (Page 88); British Gas Corporation (Page 89); British Telecom (Page 92); British Leyland Austin Rover Group Limited (Pages 97, 104); Shell UK Ltd (Page 99); Nigel Luckhurst (Page 101); Hertz Rent A Car (Page 105).

Special thanks are due to George Williams, W H Smith & Son Limited and Mrs D J Shortis of Robert Sayle, Cambridge. Line illustrations were drawn by Sarah Crompton.

EVERYDAY MATHS

ALAN LAWSON

KILBURN POLYTECHNIC

UNIVERSITY TUTORIAL PRESS

Published by University Tutorial Press Ltd.,
842 Yeovil Road, Slough SL1 4JQ.

All rights reserved—No portion of this book may be reproduced by any process without written permission from the publishers.

Published 1984

© A. Lawson 1984

0 7231 0861 7

Printed in Great Britain by J. W. Arrowsmith Ltd., Bristol BS3 2NT

SKILLS PRACTICE......

1 The Calculator

There are many different pocket calculators available. All basic calculators add, subtract, divide and multiply. Some cheap calculators also have percentage, memory and perhaps other keys.

A basic calculator is illustrated below:

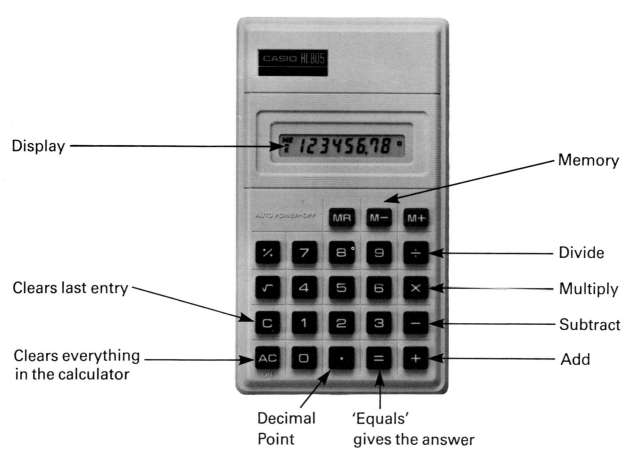

When you switch on your calculator a zero should appear on the display $\boxed{0.}$

Always remember to switch off when not using your calculator.

Calculations are carried out by pressing keys.

Examples   71 + 24

PRESS $\boxed{7}$ and $\boxed{1}$ for 71. Display reads $\boxed{71.}$

PRESS $\boxed{+}$

PRESS $\boxed{2}$ and $\boxed{4}$ for 24. Display reads $\boxed{24.}$

PRESS $\boxed{=}$ for your answer. Display reads $\boxed{95.}$

    PRESS $\boxed{AC}$ to clear the calculator before starting the next calculation.

1 The Calculator

Now try 84 − 39

Press the keys in this order:

[8] [4] [−] [3] [9] [=]   Display reads   `45.`

Remember to press [AC]

Next try 37 × 19

Press the keys in this order:

[3] [7] [×] [1] [9] [=]   Display reads   `703.`

Remember, press [AC]

Then try 1380 ÷ 15

Press the keys in this order:

[1] [3] [8] [0] [÷] [1] [5] [=]   Display reads   `92.`

Remember, press [AC]

and finally, 65 × 80 ÷ 26

Press the keys in this order:

[6] [5] [×] [8] [0] [÷] [2] [6] [=]   Display reads   `200.`

Now you try, (remember to press [AC] after each sum)

Exercise 1.1

1. 125 + 47 + 256 =
2. 540 − 359 =
3. 182 + 429 − 387 =
4. 370 × 17 =
5. 781 × 15 × 39 =
6. 1666 ÷ 49 =
7. 13260 ÷ 39 =
8. 89 × 135 ÷ 267 =
9. 84 × 19 × 120 ÷ 140 =
10. 540 × 85 × 90 ÷ 1350 =

2 Rounding

For many calculations we need to be able to 'round up' or 'round down' numbers.

Attendance figures at football matches give good examples where we can round 'up' or 'down'.

Example

25158 attended the Everton match. What is this to the nearest:
(a) thousand?   (b) hundred?

(a) 25158 is nearer to **25000** than 26000.
    (We 'round down')
(b) 25158 is nearer to 25200 than 25100.
    (We 'round up')

Exercise 2.1

1. What was the attendance at the Birmingham match to the nearest:
   (a) thousand?
   (b) hundred?
   (c) ten?

2. What was the attendance to the nearest **thousand** at:
   (a) Manchester United?
   (b) Sunderland?
   and to the nearest hundred at:
   (c) Ipswich?   (d) Tottenham?

FOOTBALL RESULTS

LEAGUE DIVISION 1

Birmingham (0) 2
Langan (pen),
Brazier
Brighton (1) 1
Foster
Everton (1) 2
King, McMahon
Ipswich (0) 0

Man. Utd (0) 1
Robson
Notts Co (2) 4
Mortimer (og)
Christie, Mair,
McCullough
Sunderland (0) 1
McCoist

Tottenham (1) 4
Crooks,
Brooke 3 (1 pen)
Watford (2) 2
Jackett, Terry

Luton (0) 3
Stein, Walsh,
Moss   13,772
Swansea (0) 1
Lewis   11,050
Man. City (1) 1
Cross   25,158
Arsenal (0) 1
Woodcock   20,792

Stoke (0) 0
                     43,132
Aston Villa (0) 1
Shaw   8,990

Southampton (0) 1
Williams   15,635

Coventry (0) 0
                     25,188

Norwich (1) 2
Deehan,
Bertschin   18,597

2 Rounding

Approximation

To obtain approximate answers to calculations we use a combination of 'rounding up' and 'rounding down'.

Example

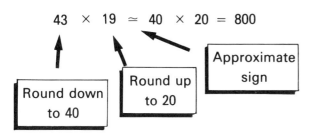

and,

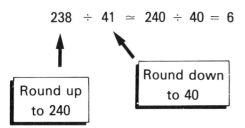

Exercise 2.2

Find approximate answers for the following:

1. 39×41

2. $182 \div 88$

3. 89×72

4. $599 \div 14$

5. $62 \times 59 \div 31$

6. $3987 \div 81$

7. 2995×52

8. The cost of six shirts at £6·97 each.

9. The cost of nine pairs of jeans at £10·99 per pair.

10. How many ties costing £1·95 could be bought with £21?

Rounding Up

For most practical activities it is necessary to round up to the nearest whole number.

Example

A man reckons he needs $6\frac{1}{2}$ rolls of wallpaper to decorate a room. He cannot buy half a roll, therefore he must buy 7 rolls.

Exercise 2.3

1. A man needs $3\frac{3}{4}$ litres of a paint which costs 95p per litre. The paint can only be bought in litres. How much does he need to spend?

2. To paint the outside of a house a decorator reckons he needs 13 litres of masonry paint.

   The paint can only be bought in 5 litre tins costing £12·25 each.
   (a) How many tins must he buy?
   (b) How much does he spend?

3. The maximum number of people allowed in a lift is 10. How many journeys must the lift make if:
   (a) 37 people, and (b) 76 people, are waiting to use it?

4. To tile their bathroom a family need 195 tiles. The tiles are sold in boxes of 20 costing £6·20 per box.
   (a) How many boxes of tiles do the family need to buy?
   (b) How much do they spend?

5. The family in question 4 intend to fit cork tiles to their bathroom floor. To tile the floor 29 tiles are required. The tiles can be bought from one shop at £2·25 for a box of four or from another shop at 70p each. How much would they save if they bought the tiles at the first shop?

3 Paying for Adverts

This shows how much it costs to place an advert in a local paper

CARPETS — from only £1·75 per square yard. You'll like our prices and our service. GM CARPETS 171 Northway Road, Northtown 61145

↓

This is a trade advert and would cost 35p per word

↓

It has the equivalent of 21 words

↓

So cost would be 21 × 35p = £7·35

SOMETHING TO SELL?
READERS PREPAID ADVERTISEMENT ORDER FORM

Readers Private Sales over £10 **25p** per word	Readers Private Cars or Homes **30p** per word
All Trade Adverts **35p** per word	Jobs Vacant or Wanted **40p** per word

Insert *one* word per square below

CARPETS	FROM	ONLY	£1.75	PER
SQUARE	YARD.	YOU'LL	LIKE	OUR
PRICES	AND	OUR	SERVICE.	G.M.
CARPETS	171	NORTHWAY	ROAD,	NORTHTOWN
61145				

Please tick box required

 private sales 25p per word   home or car 30p per word

 trade adverts 35p per word   job vacant/wanted 40p per word

Exercise 3.1 Find out how much it would cost to place the following adverts.

1. CHEST FREEZER ELECTROLUX 12 CU. FT. ONE YEAR OLD, EXCELLENT CONDITION, £50, TEL. SOUTHTOWN 59425

2. STAINLESS STEEL SINK UNIT with right hand drainer and mixer tap unit £15, Northtown 44365

3. CLOSING DOWN SALE, Three piece suites, Corner units, Dining room sets, Bean bags etc. Harry's Furniture Ltd, 215 Ridgeway Road, Southtown 71524

4. CREDA ELECTRIC COOKER £35, Electrolux Dishwasher £40, Ideal Boiler 50000 BTU £40. All good condition. Southtown 84563

5. 1976 HONDA CIVIC, MOT and taxed. An economical and reliable car £1200 o.n.o. Tel Northtown 52345

6. EXPERIENCED GARDENER wanted for permanent position with overtime available. Tel Southtown 59285

7. SALES PERSON with managerial potential for new venture opening soon in Northtown. Generous salary for right person with experience in selling electrical goods. Write with full details of experience to Box No 419 Northtown Gazette, The Lawns, Northtown.

8. HOUSE—Southtown. Reduced for quick sale. Modern detached in cul-de-sac. All amenities, cloaks, lounge, kitchen/diner, three bedrooms, bath, separate toilet, centrally heated £29 250. Southtown 65499.

4 Food: The Calorie Counter

Calories measure the energy value of food. The table below shows approximately how many calories are contained in an **average** portion of certain foods.

ITEM	CALORIES	ITEM	CALORIES
Breakfast Cereal	110	Bacon	240
Fried Egg	130	Beefsteak	390
Boiled Egg	80	Beefburger	300
Sandwiches	160	Chicken	290
Apple Pie	300	Steak & Kidney Pie	640
Milk Pudding	220	Fish	340
Cheese & Biscuits	120	Mashed Potatoes	140
White Tea/Coffee	20	Chips	495
Sugar	20 per tsp	Green Vegetables	20
Beer	200 per pt	Baked Beans	100
Chocolate Bar	250	Lamb Chop	400

Exercise 4.1

1. How many calories would be contained in a meal consisting of beefburger, chips and baked beans; apple pie; coffee, white with two sugars?

2. A couple in a restaurant order the following: One beefsteak with chips and green vegetables, one lamb chop with mashed potatoes and green vegetables, two milk puddings and two coffees each with two sugars.
How many calories have they consumed between them?

3. In a day a man has the following meals:
Breakfast—cereal, two boiled eggs, white coffee with one sugar;
Lunch—two sandwiches, one tea with sugar;
Evening meal—steak and kidney pie with chips and green vegetable, milk pudding, white coffee with one sugar.
At other times of the day he has two cups of white tea each with one sugar and two pints of beer.
How many calories has he consumed?

4. A woman on a diet needs to restrict her calorie intake to 1500 per day. One day she ate the following:
Breakfast—cereal, one boiled egg, white coffee (unsweetened);
Lunch—fish with vegetable, white coffee (unsweetened);
Evening meal—Lamb chop with mashed potatoes and vegetable, cheese and biscuits, unsweetened white coffee.
By how many calories did she under-eat or over-eat if in addition to the meals she also had three cups of unsweetened white coffee?

4 Food: The Calorie Counter

Different activities use up different amounts of energy. The chart below shows the approximate number of calories the average person uses up in carrying out certain activities.

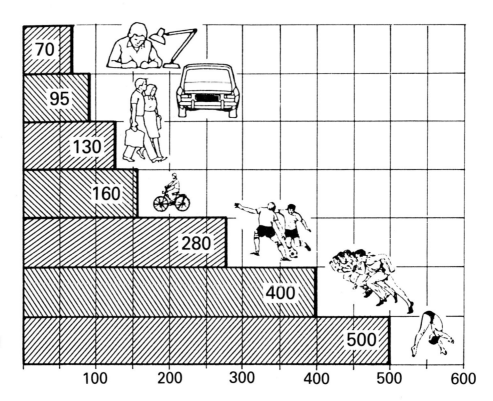

Calories used in 30 minutes →

Exercise 4.2

1. How many calories would be used up in the following:
   (a) 3 hours driving?
   (b) Playing football for 90 minutes?
   (c) Swimming for 6 minutes?
   (d) Cycling for 45 minutes?

2. How many calories would a man use if he ran for 15 minutes and then walked for 45 minutes?

3. How many calories would a boy use if he:
   studied for 3 hours, walked for 15 minutes, cycled for 30 minutes and played football for 45 minutes?

4. (a) For how long would a man need to swim to use up the energy he obtained after eating a bar of chocolate?
   (b) For how long would a boy need to run to use up the energy he obtained after eating two beefburgers?

5 Fractions (one)

Fractions are parts

This circle has been divided into six equal parts and one of the parts has been shaded. The shaded part is $\frac{1}{6}$ (one-sixth).

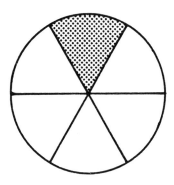

This rectangle has been divided into five equal parts and two of the parts have been shaded. The shaded part is $\frac{2}{5}$ (two fifths).

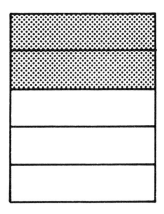

In any fraction, the **bottom** figure shows how many equal parts the whole has been divided into. The **top** figure shows how many parts are required.

The top figure is called the numerator.
The bottom figure is called the denominator.

$$\frac{5}{8} = \frac{\text{numerator}}{\text{denominator}}$$

Exercise 5.1

Write down the fraction part shaded in the figures below:

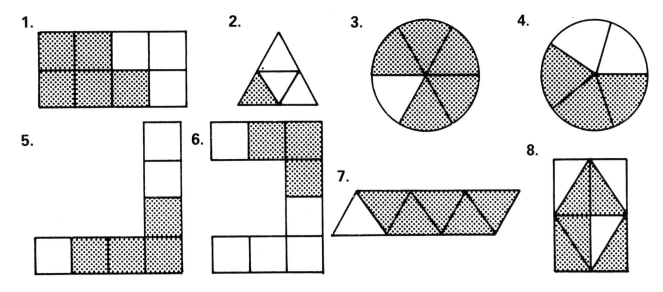

9. In the fraction $\frac{11}{12}$, what is the value of the numerator?
10. In the fraction $\frac{6}{7}$, what is the value of the denominator?

5 Fractions (one)

Improper Fractions

Improper or 'top-heavy' fractions have a numerator (top number) bigger than the denominator (bottom number). They are usually converted to give a whole number and a fraction (called a mixed number).

Example

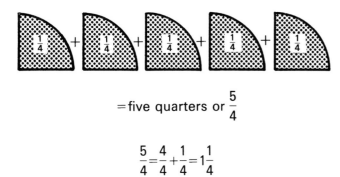

= five quarters or $\dfrac{5}{4}$

$$\dfrac{5}{4} = \dfrac{4}{4} + \dfrac{1}{4} = 1\dfrac{1}{4}$$

Exercise 5.2

Convert these improper or 'top-heavy' fractions to mixed numbers.

1. $\dfrac{3}{2}$ 2. $\dfrac{7}{5}$ 3. $\dfrac{8}{7}$ 4. $\dfrac{15}{8}$

5. $\dfrac{15}{2}$ 6. $\dfrac{22}{7}$ 7. $\dfrac{31}{15}$ 8. $\dfrac{53}{10}$

9. $\dfrac{117}{100}$ 10. $\dfrac{55}{3}$

For some calculations, mixed numbers need to be converted back into improper fractions.

Example $1\dfrac{4}{5} = \dfrac{5}{5} + \dfrac{4}{5} = \dfrac{9}{5}$

Exercise 5.3

Convert the following mixed numbers back to improper fractions:

1. $1\dfrac{1}{2}$ 2. $1\dfrac{4}{5}$ 3. $2\dfrac{1}{5}$ 4. $1\dfrac{9}{10}$

5. $1\dfrac{7}{8}$ 6. $2\dfrac{3}{4}$ 7. $3\dfrac{1}{2}$ 8. $4\dfrac{2}{3}$

9. $7\dfrac{1}{7}$ 10. $16\dfrac{2}{3}$

Cancelling Down

Some fractions can be made simpler by cancelling down. Cancelling is carried out by dividing top and bottom by the same number.

Examples

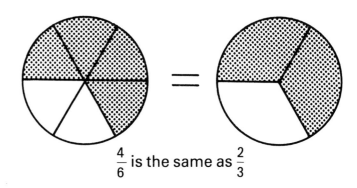

$\dfrac{4}{6}$ is the same as $\dfrac{2}{3}$

$\dfrac{4 \div 2}{6 \div 2}$ | Divide top and bottom by 2 | $= \dfrac{2}{3}$

Exercise 5.4

Cancel down the following:

1. $\dfrac{5}{10}$ 2. $\dfrac{16}{20}$ 3. $\dfrac{8}{12}$ 4. $\dfrac{12}{16}$

5. $\dfrac{10}{24}$ 6. $\dfrac{8}{32}$ 7. $\dfrac{36}{48}$ 8. $\dfrac{25}{75}$

9. $\dfrac{20}{50}$ 10. $\dfrac{100}{150}$ 11. $\dfrac{200}{250}$ 12. $\dfrac{250}{600}$

13. $\dfrac{17}{34}$ 14. $\dfrac{13}{39}$ 15. $\dfrac{125}{200}$ 16. $\dfrac{15}{9}$

17. $\dfrac{35}{20}$ 18. $\dfrac{60}{15}$

19. A factory employs 2000 workers. Of this number 800 are female.
    (a) What fraction of the workers are female?
    (b) What fraction of the workers are male?

20. A firm employing 45 workers faces a total wage bill of £5400 each week. How much on average does each worker receive?

6 Fractions (two)

Equivalent Fractions

For some calculations we need to find different ways of writing the same fraction.

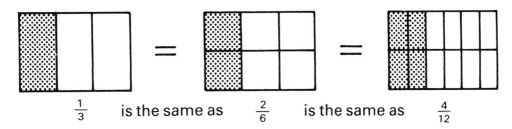

$\frac{1}{3}$ is the same as $\frac{2}{6}$ is the same as $\frac{4}{12}$

To make fractions into equivalents we need to multiply top and bottom by the same number:

Example

$$\frac{2}{3} \xrightarrow{\times 2} \frac{4}{6} \quad \text{and} \quad \frac{4}{5} \xrightarrow{\times 3} \frac{12}{15}$$

Exercise 6.1

Complete the following equivalent fractions:

1. $\frac{1}{2} = \frac{}{4}$
2. $\frac{2}{5} = \frac{}{10}$
3. $\frac{1}{10} = \frac{2}{}$
4. $\frac{7}{10} = \frac{70}{}$
5. $\frac{1}{3} = \frac{}{12}$
6. $\frac{2}{3} = \frac{6}{}$
7. $\frac{6}{7} = \frac{18}{}$
8. $\frac{3}{8} = \frac{6}{}$
9. $\frac{5}{6} = \frac{15}{}$
10. $\frac{7}{12} = \frac{28}{}$
11. $\frac{9}{10} = \frac{}{100}$
12. $\frac{9}{14} = \frac{}{42}$
13. $\frac{2}{3} = \frac{}{36}$
14. $\frac{57}{9} = \frac{}{81}$
15. $\frac{7}{12} = \frac{84}{}$
16. $\frac{11}{13} = \frac{}{169}$
17. $\frac{4}{3} = \frac{}{12}$
18. $\frac{17}{4} = \frac{68}{}$

19. In a survey carried out before a local election 240 people were questioned. $\frac{3}{8}$ said they would vote Labour and $\frac{7}{24}$ said they would vote Conservative.
    (a) How many said they would vote Labour? i.e. $\frac{3}{8} = \frac{?}{240}$
    (b) How many said they would vote Conservative?
    (c) How many **more** said they would vote Labour than Conservative?

20. A firm manufacturing model ships makes them to a size equivalent to $\frac{1}{100}$th of the original ships. What would be the real sizes of:
    (a) The mast of a sailing ship measuring 10 cms on the model i.e. $\frac{1}{100} = \frac{10}{?}$ (Give your answer in metres)
    (b) The length of a liner measuring 35 cm on the model. (Answer in metres).
    (c) A ships life boat measuring 3 cm on a model. (Answer in metres).

6 Fractions (two)

Addition and Subtraction of Fractions

If the denominator is the same:

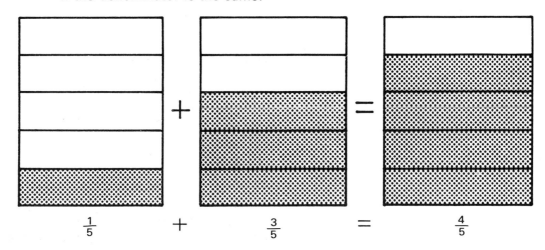

$$\frac{1}{5} + \frac{3}{5} = \frac{4}{5}$$

We add together

If the denominator is different:

Example 1 $\quad \frac{1}{2} + \frac{1}{3}$ Make into equivalent fractions with the same denominator

$$\frac{3}{6} + \frac{2}{6} = \frac{5}{6}$$

Example 2 $\quad \frac{5}{6} - \frac{1}{4}$ Make into equivalent fractions with the same denominator

$$\frac{10}{12} - \frac{3}{12} = \frac{7}{12}$$

Exercise 6.2

Cancel down where necessary.

1. $\frac{1}{3} + \frac{1}{3}$
2. $\frac{5}{6} - \frac{2}{6}$
3. $\frac{1}{3} + \frac{1}{6}$
4. $\frac{2}{3} - \frac{1}{2}$
5. $\frac{1}{2} + \frac{3}{8}$
6. $\frac{2}{5} + \frac{3}{10}$
7. $\frac{5}{12} - \frac{1}{3}$
8. $1\frac{1}{2} + 1\frac{1}{4}$
9. $2\frac{1}{2} + \frac{5}{12}$
10. $\frac{11}{24} - \frac{1}{3}$
11. $\frac{19}{20} - \frac{13}{40}$
12. $\frac{99}{100} - \frac{19}{20}$

6 Fractions (two)

Subtraction involving mixed numbers:

Example

$$1\frac{1}{3} - \frac{5}{6}$$
$$\downarrow \quad \downarrow$$
$$\frac{4}{3} - \frac{5}{6}$$
$$\downarrow \quad \downarrow$$
$$\frac{8}{6} - \frac{5}{6} = \frac{3}{6} = \frac{1}{2}$$

Make $1\frac{1}{3}$ into improper fraction

Then make into equivalent fractions with the same denominator

13. $1\frac{5}{8} - \frac{11}{16}$

14. $2\frac{1}{2} - 1\frac{7}{8}$

15. $3\frac{1}{4} - 2\frac{11}{12}$

16. $5\frac{1}{6} - 4\frac{7}{12}$

17. A disc jockey plays two records. The first lasts $2\frac{5}{8}$ minutes and the second lasts $3\frac{1}{4}$ minutes. Find the total time taken to play the two records.

18. To mix paint $1\frac{3}{4}$ litres of one colour is added to $2\frac{2}{5}$ litres of another colour. How much of the mixture is there now?

19. A man waited $8\frac{3}{4}$ minutes for a bus and $6\frac{3}{4}$ minutes travelling on the bus.
    (a) How long did his journey take?
    (b) How much longer did he spend waiting than actually travelling on the bus?

BBC Copyright Photograph

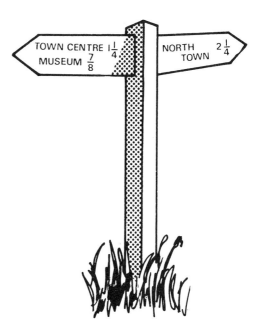

TOWN CENTRE $1\frac{1}{4}$
MUSEUM $\frac{7}{8}$
NORTH TOWN $2\frac{1}{4}$

20. (a) How far from Northtown to the town centre?
    (b) How far from Northtown to the museum?
    (c) How much further to the town centre than to the museum?

7 Fractions (three)

Multiplication

Examples:

1. [diagram] $\frac{1}{3} \times 4 = \frac{4}{3}$ or $1\frac{1}{3}$

   or $\frac{1}{3} \times \frac{4}{1} = \frac{4}{3} = 1\frac{1}{3}$

2. [diagram] $\frac{3}{4} \times \frac{1}{3} = \frac{1}{4}$

   or $\frac{3^1}{4} \times \frac{1}{3_1} = \frac{1}{4}$

3. $\frac{4}{5}$ of $\frac{2}{3}$ ← multiply tops together and bottoms together → of means the same as times

   $\frac{4 \times 2}{5 \times 3} = \frac{8}{15}$

Exercise 7.1

Cancel down where necessary.

1. $\frac{1}{3} \times 6$
2. $\frac{1}{4} \times 7$
3. $\frac{2}{5}$ of 3
4. $\frac{3}{4} \times 12$
5. $\frac{3}{8} \times \frac{1}{2}$
6. $\frac{4}{5} \times \frac{1}{4}$
7. $\frac{7}{8} \times \frac{13}{14}$
8. $\frac{9}{10}$ of $\frac{10}{12}$
9. $\frac{15}{16} \times \frac{4}{5}$
10. $\frac{8}{9} \times \frac{9}{10}$
11. $\frac{19}{20} \times \frac{5}{38}$
12. $\frac{4}{7}$ of $\frac{14}{15}$

When we have mixed numbers:

Example: $1\frac{1}{2} \times 1\frac{3}{4}$ ← Convert to improper fractions

$\frac{3}{2} \times \frac{7}{4} = \frac{3 \times 7}{2 \times 4} = \frac{21}{8} = 2\frac{5}{8}$

13. $1\frac{3}{5} \times \frac{3}{4}$
14. $1\frac{1}{4}$ of $1\frac{1}{5}$
15. $2\frac{1}{2} \times \frac{8}{9}$
16. $3\frac{1}{2} \times 1\frac{1}{3}$

7 Fractions (three)

17. Find (a) How many minutes in $\frac{2}{3}$ of 2 hours?
    and (b) How many seconds in $\frac{3}{4}$ of 2 minutes?

18. A boy walks at a speed of $5\frac{1}{2}$ kilometres per hour. How far will he walk in
    (a) $\frac{1}{2}$ hour?   (b) $1\frac{1}{2}$ hours?

19. This gauge is for the petrol tank of a car. The tank holds $10\frac{1}{2}$ gallons when full.

    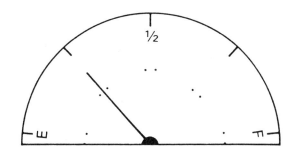

    (a) How much petrol is in the tank now?
    (b) How much petrol would be needed to fill the tank up?
    (c) How much petrol would be in the tank if the gauge indicated $\frac{1}{3}$ full?

20. At a pop concert lasting $2\frac{1}{2}$ hours, three groups shared the time out as follows:

    First group $\frac{1}{5}$ of the total time.
    Second group $\frac{1}{3}$ of the total time.
    Third group remainder of the time.

    (a) How long did the first group play for?
    (b) How long did the second group play for?
    (c) What fraction of the total time did the third group play for?

Division of Fractions

Examples

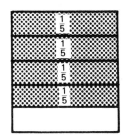 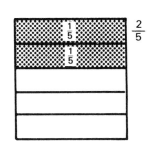

and, $1\frac{1}{3} \div \frac{1}{3}$ means

 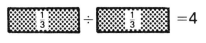

RULE: When dividing by fractions change the ÷ sign to × and turn the divisor upside down.

Example

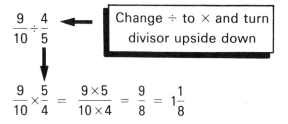

$$\frac{9}{10} \times \frac{5}{4} = \frac{9 \times 5}{10 \times 4} = \frac{9}{8} = 1\frac{1}{8}$$

Exercise 7.2
Cancel down where necessary.

1. $4 \div \frac{1}{5}$
2. $3 \div \frac{3}{5}$
3. $\frac{1}{2} \div \frac{1}{4}$

4. $\frac{9}{10} \div \frac{4}{5}$
5. $\frac{4}{5} \div \frac{1}{3}$
6. $\frac{5}{8} \div \frac{1}{4}$

7. $\frac{5}{16} \div \frac{7}{8}$
8. $\frac{3}{4} \div \frac{7}{10}$
9. $\frac{5}{6} \div \frac{11}{12}$

10. $\frac{11}{24} \div \frac{55}{96}$
11. $\frac{15}{16} \div \frac{5}{6}$
12. $\frac{3}{4} \div \frac{16}{17}$

7 Fractions (three)

When we have mixed numbers:

Example

$$2\tfrac{1}{2} \div 1\tfrac{1}{9}$$

Make into improper fractions

$$\tfrac{5}{2} \div \tfrac{10}{9}$$

Change ÷ to × and turn divisor upside down

$$\tfrac{5}{2} \times \tfrac{9}{10} = \tfrac{5 \times 9}{2 \times 10} = \tfrac{9}{4} = 2\tfrac{1}{4}$$

13. $2\tfrac{1}{3} \div 1\tfrac{1}{6}$ 14. $4\tfrac{1}{4} \div 1\tfrac{1}{3}$

15. $\tfrac{5}{8} \div 1\tfrac{1}{4}$ 16. $2\tfrac{3}{11} \div \tfrac{10}{11}$

17. (a) $2\tfrac{1}{2}$ litres of paint cost £3.75. How much for 1 litre?
(b) A boy walks $12\tfrac{1}{2}$ miles in $1\tfrac{1}{4}$ hours. How many miles per hour does this represent?

18. (a) A machine takes $\tfrac{1}{5}$ hour to make a part. How many items will the machine make in $3\tfrac{1}{5}$ hours?
(b) How many books each $\tfrac{3}{4}$ cm thick could be stacked in a box 21 cm high?

19. (a) A piece of metal $2\tfrac{1}{5}$ metres in length is to be cut into 22 equal lengths. How long will each length be?
(b) The bus journey round a city centre takes $\tfrac{1}{8}$ hour to complete. How many journeys in $2\tfrac{1}{4}$ hours?

20. (a) The average length of time taken to play a pop record is $3\tfrac{3}{4}$ minutes. How many could be played in $1\tfrac{1}{4}$ hours?
(b) How many books $1\tfrac{1}{4}$ cm thick could be stacked side by side on a book shelf 225 cm long?

Calculator Work

The calculator can be used to solve some problems where fractions are involved.

Example   Find $\tfrac{2}{5}$ of 145

$\tfrac{2}{5}$ of 145 is the same as $\tfrac{2}{5} \times \tfrac{145}{1} = \tfrac{2 \times 145}{5}$

Press keys

   Display reads   58

Exercise 7.3

Find the value of:

1. $\tfrac{3}{7}$ of 1470   2. $\tfrac{1}{9}$ of 1809   3. $\tfrac{5}{12}$ of 2880   4. $\tfrac{11}{12}$ of 504

5. $\tfrac{13}{14}$ of 700   6. $\tfrac{11}{34} \times 1020$ 7. $\tfrac{5}{18}$ of £5490   8. $\tfrac{5}{17}$ of £2980

9. $\tfrac{7}{12}$ of $1512 \times \tfrac{1}{2}$ 10. $\tfrac{4}{5}$ of $1800 \times \tfrac{1}{4}$

8 Decimals

Fraction parts with denominators of 10 and 100 can be written in decimal form.

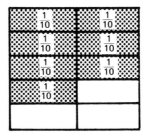

$\frac{7}{10}$ can be written as 0·7

$1\frac{9}{10}$ can be written 1·9

$\frac{23}{100}$ can be written 0·23

and $2\frac{7}{100}$ can be written 2·07

Exercise 8.1

Write out the following in decimal form:

1. $1\frac{3}{10}$ 2. $\frac{73}{100}$ 3. $2\frac{6}{10}$ 4. $3\frac{52}{100}$ 5. $1\frac{17}{100}$

6. $\frac{4}{10}$ 7. $7\frac{45}{100}$ 8. $\frac{12}{100}$ 9. $\frac{99}{100}$ 10. $1\frac{84}{100}$

MONEY IS AN EVERYDAY EXAMPLE IN WHICH DECIMAL FORM IS USED

Examples £$2\frac{42}{100}$ = £2·42

But £$1\frac{5}{10}$ = £1·50

and £$1\frac{7}{100}$ = £1·07

Two decimal places must always be used when dealing with money

Exercise 8.2

Write out the following amounts in decimal form:

1. £$\frac{17}{100}$ 2. £$1\frac{3}{10}$ 3. £$1\frac{11}{100}$ 4. $4\frac{1}{100}$ 5. £$6\frac{6}{10}$

6. £$5\frac{1}{10}$ 7. £$\frac{53}{100}$ 8. £$\frac{3}{100}$ 9. $6\frac{65}{100}$ 10. $1\frac{2}{100}$

8 Decimals

Addition and Subtraction of Decimals

When adding and subtracting decimals, write them out in the usual way lining up the decimal points underneath each other.

Examples   3·2 + 9·7   and   5·7 − 4·5

```
  3·2            5·7
+ 9·7          − 4·5
─────          ─────
 12·9            1·2
```

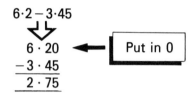

```
  6·20
− 3·45
──────
  2·75
```
← Put in 0

Exercise 8.3

Complete the following:

1. 0·9 + 3·7
2. 1·1 + 6·2 + 0·9
3. 1·9 − 0·7
4. 2·45 − 1·29
5. 10·9 + 5·68
6. 2·4 + 3·56 + 4·65
7. 4·5 − 3·65
8. 11·9 − 9
9. 1·06 + 1·9 + 3·73
10. 3·7 + 1·12 + 4·57
11. 3·43 − 0·62
12. 5·9 − 3·98
13. 6·79 + 5·4 − 8·6
14. 11·04 + 5·92 − 13·8
15. 1·01 + 11 − 9·89
16. 15·6 + 18·54 − 29·86
17. 14 + 16·85 − 25·49
18. 18·2 + 19·4 − 35·85

19. (a) How much would I spend if I bought items costing £0·43, 59p, £1·56 and 64p?
    (b) How much change would I receive from £5?

20. Convert to decimal form and add together

    $\frac{16}{100}$, $\frac{9}{10}$, $1\frac{31}{100}$ and $\frac{54}{100}$

Multiplication of Decimals by 10 or 100

Examples

1. 6·2 × 10 = 6·2̂ = 62
    and 7·25 × 10 = 7·2̂5 = 72·5

2. 7·19 × 100 = 7·1̂9̂ = 719
    and 6·432 × 100 = 6·4̂3̂2 = 643·2

> WHEN MULTIPLYING BY 10 OR 100, THE DECIMAL POINT MOVES 1 OR 2 PLACES TO THE RIGHT

Exercise 8.4

Complete the following:

1. 1·9 × 10 =
2. 0·25 × 10 =
3. 0·62 × 10 =
4. 9·97 × 10 =
5. 8·375 × 100 =
6. 6·075 × 100 =

7. How much for 10 ties costing £1.65 each?

8. 4·465 × 100 =
9. 2·006 × 1000 =

10. How much for 100 pens costing £0·12 each?

Multiplication of Decimals by Numbers other than 10 or 100 etc.

Examples

1·2 × 11   [Forget decimal part and do 12 × 11]

12 × 11 = 132   [How many decimal places in the question?]

One in this case, so put **one** in the answer by counting in one from the right hand end.

So, **132** becomes **13·2**

8 Decimals

and,

1·6 × 0·9

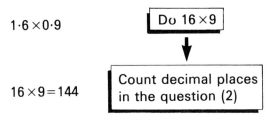

16 × 9 = 144

So, **144** becomes **1·44**

Exercise 8.5

Write out and complete the following:

1. 1·9 × 9
2. 2·4 × 11
3. 1·2 × 0·7
4. 35·9 × 8
5. 3·4 × 0·6
6. 1·1 × 1·2
7. 16·4 × 0·2
8. 311 × 0·12

9. A firm pays part-time workers at £1·22 per hour. How much would they pay for:
   (a) 7 hours
   (b) 11 hours

10. (a) A bicycle was priced at £80. During a sale it was sold at 0·75 times this amount. What was its selling price?
    (b) How far does a man walk in 2·5 hours at a speed of 4·8 mph?

Division by 10 or 100

Examples

1. $\quad$ 12·5 ÷ 10 = 12·5 = 1·25

   and 11·09 ÷ 10 = 11·09 = 1·109

2. $\quad$ 9120·5 ÷ 100 = 120·5 = 1·205

   and 16·7 ÷ 100 = 16·7 = 0·167

> WHEN DIVIDING BY 10 OR 100 THE DECIMAL POINT MOVES 1 OR 2 PLACES TO THE LEFT

Exercise 8.6

Write out and complete the following:

1. 2·7 ÷ 10
2. 124 ÷ 100
1. 5·6 ÷ 10
4. 11·5 ÷ 10
5. 1·9 ÷ 100
6. £0·60 ÷ 10

7. Ten pairs of socks cost £6·60. Find the cost of one pair.

8. 1·6 ÷ 100
9. 62 ÷ 100

10. One hundred items cost £7·00. Find the cost of one.

Division of Decimals by Numbers other than 10 or 100 etc.

Examples

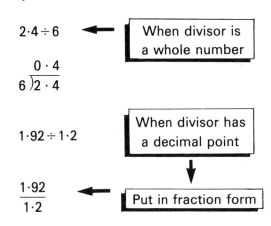

> When the divisor is a decimal make it into a whole number by moving the decimal point. **But remember** to move the decimal point in the numerator (top) the same number of places.

8 Decimals

Exercise 8.7

Write out and complete the following:

1. $2.88 \div 12$
2. $14 \div 4$
3. $1.32 \div 1.1$
4. $81.9 \div 0.9$
5. $9 \div 1.5$
6. $0.55 \div 0.2$
7. $0.123 \div 0.3$
8. $11.5 \div 2.3$

9. An electric bill was £72·54 for 1 quarter (13 weeks). How much is this per week?

10. A mans rate of pay is £2·20 per hour. How many hours did he work for £90·20?

9 Changing Fractions to Decimals

To change fractions to decimals we divide the bottom into the top.

Fractions that Recur

Examples

1. Change $\frac{1}{4}$ to decimal form

   Divide bottom into top

   $$\begin{array}{r} 0\cdot 25 \\ 4\overline{)1\cdot 00} \end{array}$$ ← Write the top as a decimal if necessary

   So, $\frac{1}{4} = 0\cdot 25$

2. Change $\frac{3}{8}$ to decimal form

   $$\begin{array}{r} 0\cdot 375 \\ 8\overline{)3\cdot 000} \end{array}$$ ← Write the top as a decimal if necessary

Example

Change $\frac{1}{3}$ to decimal form

$$\begin{array}{r} 0\cdot 3333 \\ 3\overline{)1\cdot 0000} \end{array}$$ ← The 3 keeps recurring

So, $\frac{1}{3} = 0\cdot \dot{3}$ The dot above the 3 indicates recurring decimal

Fractions that recur or don't work out are best checked using a calculator.

Example

Change $\frac{1}{6}$ to decimal form

Press keys

[1] [÷] [6] [=]   Display reads

$$\boxed{0.1666666}$$

So, $\frac{1}{6} = 0\cdot 1\dot{6}$ ← The dot indicates recurring decimal

Exercise 9.1

Change the following into decimal form:

1. $\frac{1}{2}$
2. $\frac{3}{4}$
3. $\frac{2}{5}$
4. $\frac{1}{8}$
5. $\frac{7}{8}$
6. $\frac{4}{5}$
7. $\frac{1}{20}$
8. $\frac{11}{20}$
9. $\frac{1}{25}$
10. $\frac{4}{25}$

Exercise 9.2

Use your calculator to change the following into decimal form and indicate which of them are recurring.

1. $\frac{2}{3}$
2. $\frac{5}{8}$
3. $\frac{19}{20}$
4. $\frac{5}{6}$
5. $\frac{1}{9}$
6. $\frac{5}{16}$
7. $\frac{5}{9}$
8. $\frac{7}{9}$
9. $\frac{11}{12}$
10. $\frac{7}{32}$

9 Changing Fractions to Decimals

Other Decimals that Don't Work Out

Example 1   Change $\frac{3}{7}$ to decimal form

Press keys

[3] [÷] [7] [=]   Display reads 0.4285714

The numbers **neither** end nor recur

With decimals like that we can round off the number to **2 decimal places (d.p.)**

Display reads 0.4285714

If the third decimal place is 5 or greater we add one to the second decimal place.

So, $\frac{3}{7} = 0.43$ (to 2 decimal places)

Example 2   Change $\frac{2}{13}$ to decimal form

Press keys

[2] [÷] [1] [3] [=]   Display reads 0.1538461

If the third decimal place is less than 5 we leave the second decimal place alone

So, $\frac{2}{13} = 0.15$ (to 2 decimal places)

Exercise 9.3

Use your calculator to change the following fractions to decimals giving your answer to **two** decimal places.

1. $\frac{4}{23}$
2. $\frac{9}{13}$
3. $\frac{12}{17}$
4. $\frac{1}{7}$
5. $\frac{2}{7}$

6. $\frac{45}{71}$
7. $\frac{56}{59}$
8. $\frac{23}{78}$
9. $\frac{12}{13}$
10. $\frac{129}{145}$

10 Percentages

A percentage is a fraction with a **denominator** of 100.

The percentage symbol '%' means out of 100.

Changing Percentages to Fractions

Example Change 40% to a fraction.

$$40\% = \frac{40}{100} \quad \leftarrow \text{We can cancel down}$$

$$\frac{40}{100} = \frac{2}{5}$$

So, $40\% = \frac{2}{5}$

Exercise 10.1 Change to fractions, cancelling down where possible.

1. 25%   2. 50%   3. 65%   4. 80%
5. 85%   6. 16%   7. 27%   8. 52%
9. 36%   10. 49%

Changing Fractions to Percentages

Example Change $\frac{4}{5}$ to a percentage.

$$\frac{4}{5} \times \frac{100}{1} = \frac{4 \times 100}{5 \times 1} \quad \leftarrow \text{Cancel down}$$

$$= 80\%$$

To change any fraction to a percentage multiply by 100

Exercise 10.2. Change the following to percentages.

1. $\frac{3}{4}$ 2. $\frac{9}{10}$ 3. $\frac{1}{20}$ 4. $\frac{1}{3}$

5. $\frac{2}{3}$ 6. $\frac{11}{20}$ 7. $\frac{1}{5}$

8. $\frac{3}{5}$ of the students at a college are female.
   (a) What percentage is this?
   (b) What percentage are male students?

9. A firm exports $\frac{9}{20}$ of the products they manufacture.

   (a) What percentage is this?
   (b) What percentage do they sell at home?

10. In a survey on local traffic it was found that $\frac{1}{5}$ of the vehicles were driven by women, $\frac{1}{20}$ of the vehicles were vans and $\frac{3}{10}$ of the vehicles were of foreign manufacture.
    (a) What percentage were driven by women?
    (b) What percentage were vans?
    (c) What percentage were of foreign manufacture?

10 Percentages

IMPORTANT PERCENTAGE-FRACTION EQUIVALENTS TO REMEMBER ARE

$50\% = \frac{1}{2}$

$25\% = \frac{1}{4}$

$75\% = \frac{3}{4}$

$33\frac{1}{3}\% = \frac{1}{3}$

$66\frac{2}{3}\% = \frac{2}{3}$

Percentage Parts of a Quantity

Example Find 70% of £50.

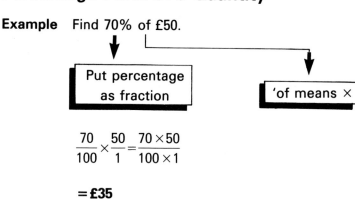

$\frac{70}{100} \times \frac{50}{1} = \frac{70 \times 50}{100 \times 1}$

$= £35$

Exercise 10.3 Complete the following using the fraction equivalents where possible

1. 30% of 120
2. 20% of 60 miles
3. 44% of £250
4. $33\frac{1}{3}\%$ of £420
5. 16% of 50
6. $66\frac{2}{3}\%$ of £150
7. 35% of 160 seconds
8. $12\frac{1}{2}\%$ of £640

9. Workers earning £80 per week are given a 10% wage increase.
   (a) What is the increase amount?
   (b) What will be their new weekly wage?

10. The price of a colour TV set is to be increased by 15%. If the original price was £240 find
    (a) The increase amount.
    (b) The new price.

Calculator Work

More difficult percentage parts can be found with a calculator.

Example Find 7·6% of £95.

Press keys

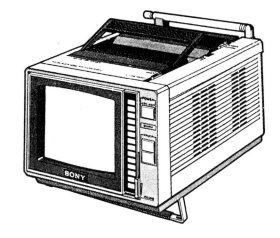

Display reads 7.22

This means that 7·6% of £95 = £7·22

Exercise 10.4 Use your calculator to find:

1. 8·5% of £110
2. 12% of 117
3. 78% of £124
4. 2·5% of 84
5. 12·5% of £112·64
6. 19% of 361
7. 3·9% of £60
8. 54% of 108
9. 11·3% of 120
10. 9·75% of 850

10 Percentages

Increase and Decrease

We can use the calculator to increase or decrease quantities by given percentages. This is useful when we need to **MARK UP** (increase) or give **DISCOUNT** (decrease) on prices.

Examples

1. Mark up (increase) £24 by 15%

Press keys

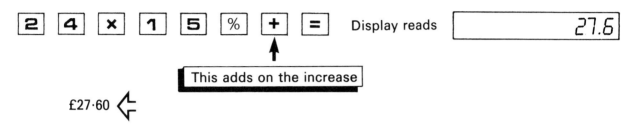

This adds on the increase

£27·60

2. Give a discount of 18% on £120

Press keys

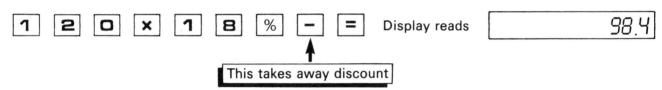

This takes away discount

The reduced price is therefore £98·40

Exercise 10.5

1. **Mark up** the following:
   (a) £12·50 by 50%   (b) £4200 by 8·5%

2. Give the following **discounts**:
   (a) 12% on £560   (b) 25% on £625

3. A car manufacturer decides to **increase** all new car prices by 7·2%.
   What will be the new marked up prices on models at:
   (a) £4950?   (b) £6250?

4. During a sale a shop allows 30% **discount** on the marked price of all goods. How much would you pay for (a) a colour TV marked at £350, and (b) a music centre at £195?

5. A firm decides to give all their workers a 6·5% wage increase. What will be the new weekly wage for workers earning:
   (a) £92 per week?   (b) £132 per week?

6. If a bill for £1240 is paid within a month a discount of 7·5% is offered. How much will you have to pay if you paid the bill within one month?

7. A shopkeeper decides to mark up certain prices by 8%. Find to the nearest penny the new prices on items marked at:
   (a) £1·55   (b) 96p.

8. A householders' average quarterly gas bill is £105. What will be the new quarterly average if prices rise by 6·5%?

9. World wide sales of a pop group's first disc amounted to 2,500,000. Their second disc showed a decrease of 47% on this figure. Find the sales figure for the second disc.

10. The population of a small country is 2,000,000. Its population is expected to rise by 9·5% each year. Find its population in
    (a) 1 year (b) 2 years.

11 Weights and Measures

Metric Measures

Metres, grams and litres are the basic units for measuring **length**, **weight** and **capacity**.
Metric measurement is based on tens, hundreds and thousands.

LENGTHS

10 millimetres (mm) = 1 centimetre (cm)
100 centimetres (cm) = 1 metre (m)
1000 metres (m) = 1 kilometre (km)

WEIGHT

1000 grams (g) = 1 kilogram (kg)

CAPACITY

1000 millilitres (ml) = 1 litre (l)

When possible use the information in the table to answer the following questions.

1. Measure the lines shown below in cm and in mm. The first one is 4·5 cm or 45 mm.
   (a)

   W _____
   X _____
   Y _____
   Z _____

   (b) What would be the length of a line three times as long as X in mm.

2. 152 cm can be written 1·52 m. Write in metres
   (a) 247 cm   (b) 75 cm   (c) 9 cm

3. How many cm in:
   (a) $\frac{1}{2}$ m?   (b) $\frac{3}{4}$ m?   (c) $\frac{2}{5}$ m?

4. Add together 1·17 m, 256 cm and 1247 mm. Give your answer in metres.

5. What fraction of 1 kg is:
   (a) 500 g?   (b) 300 g?

6. The weights of three men are 70 kg, 65 kg and 66 kg. Find their average weight.

7. Steak costs £2·50 per kg. Find the cost of
   (a) $1\frac{1}{2}$ kg   (b) 200 g

8. How many millilitres in:
   (a) $\frac{1}{4}$ litre?   (b) $\frac{1}{10}$ litre?

9. What fraction of 1 litre is:
   (a) 200 ml?   (b) 500 ml?

10. 5 ml spoons are used to measure medicine. How many spoonfuls in:
    (a) 1 litre?   (b) 350 ml?

11. The length of a swimming pool is 50 metres. How many lengths must be swum to complete
    (a) 1.5 km?   (b) 5 km?

12. Included amongst the ingredients used by a bakery to make a pie are 400 grams of minced beef and 150 grams of plain flour.
    (a) How many kg of minced beef would be required for 25 pies?
    (b) How many pies could be made with 4·5 kg of plain flour?

13. An orange drink is made up with 25 ml of concentrated orange and 225 ml of water.
    (a) How much water should be mixed with 16 ml of orange?
    (b) How much orange should be mixed with 900 ml of water?

14. 1 litre of water weighs 1 kg. What would be the weight of
    (a) 200 ml
    (b) 350 ml
    (c) 15 ml of water?
    and
    (d) What capacity would be represented by $1\frac{3}{4}$ kg of water?

11 Weights and Measures

Imperial Measures

In the imperial systems inches, feet, yards and miles are used for measuring **lengths,** ounces, pounds, and stones for **weight,** and pints and gallons for **capacity.**

Use the information in the table to answer the following questions.

LENGTH
- 12 inches (in) = 1 foor (ft)
- 3 feet (ft) = 1 yard (yd)
- 1760 yards (yd) = 1 mile

WEIGHT
- 16 ounces (oz) = 1 pound (lb)
- 14 pounds (lbs) = 1 stone (st)

CAPACITY
- 8 pints (pts) = 1 gallon

Exercise 11.2

1. How many inches in:
   (a) $\frac{1}{3}$ of a foot (b) $\frac{1}{2}$ a yard
   (c) 1 yard, 1 foot and 1 inch?

2. How many ounces in:
   (a) $\frac{3}{4}$ lb (b) $\frac{5}{8}$ lb (c) $3\frac{1}{2}$ lb?

3. How many pints in:
   (a) $\frac{3}{4}$ gallon (b) $1\frac{3}{8}$ gallons (c) $5\frac{7}{8}$ gallons?

4. How many yards in:
   (a) $\frac{1}{2}$ mile (b) $\frac{1}{8}$ mile (c) $\frac{1}{10}$ mile?

5. A man is 5 ft $7\frac{1}{2}$ in tall, his daughter is 4 ft 9 in. Find the difference in their heights.

6. An athlete drinks 2 pints of milk every day. How many gallons will the athlete drink in 12 weeks?

7. For a birthday party a girl intends to bake 3 cakes. The cakes need $\frac{1}{2}$ lb, $\frac{3}{4}$ lb and $\frac{7}{8}$ lb of flour respectively. How many ounces of flour are required?

8. Timber is sold by the foot. How much would it cost for:
   (a) $5\frac{1}{2}$ feet at 65p per foot?
   (b) 8 feet at $49\frac{1}{2}$p per foot?

9. (a) A man weighs $12\frac{1}{2}$ stones, his wife weighs 8 stones 11 pounds. By how many pounds is the man heavier than his wife?
   (b) A heavyweight boxer weighs 224 lbs. How many stones is this?

10. The petrol tank of this car holds 12 gallons when full. The petrol consumption of the car is 40 miles per gallon.
    (a) How far would the car travel if the tank was $\frac{3}{4}$ full?
    (b) How many gallons would be used if the car travelled 380 miles?

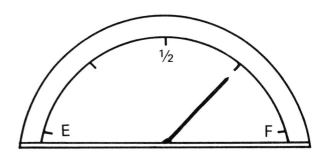

11 Weights and Measures

Conversions

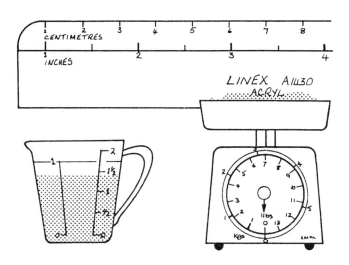

IMPERIAL—METRIC	METRIC—IMPERIAL
Approximately	Approximately
1 inch — $2\frac{1}{2}$ cm	1 cm = 0·4 in
1 foot = 30 cm	1 m = 40 in
1 mile = 1·6 km	1 km = $\frac{5}{8}$ mile
1 pound (lb) = 450 g	1 kg = 2·2 lbs

Exercise 11.3

Use the information in the tables to answer the following questions.

1. (a) How many cm in 5 inches?
   (b) How many km in 15 miles?
   (c) How many kg in 20 lbs?

2. (a) How many inches in 12 cm?
   (b) How many miles in 64 km?
   (c) How many lbs in 50 kg?

3. A pair of trousers is marked 32 inch waist. What is this measurement in cm?

4. What is the chest size in inches of a sweater marked at 105 cm?

5. A man is 1·8 metres tall. What is his height in feet?

The diagram below compares Celsius and Fahrenheit temperatures. Use your ruler to answer the following questions.

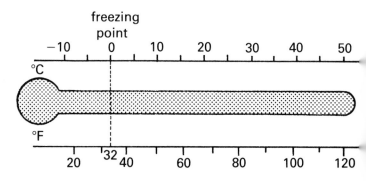

6. Convert the following:
   (a) 80° Fahrenheit to Celsius.
   (b) 10° Celsius to Fahrenheit.
   (c) 40° Fahrenheit to Celsius.

7. (a) Normal body temperature is 37° Celsius. What is this in Fahrenheit?
   (b) 30° Celsius is very hot weather. What is this in Fahrenheit?
   (c) 55° Fahrenheit is mild weather. What is this in Celsius?

11 Weights and Measures

METRIC CONVERSION CHART					
Litres	Galls	Litres	Galls	Price	
				Per Lit.	Per Gall
1	·22	42	9·24		
2	·44	44	9·68	29p	131·9p
4	·88	46	10·12	30p	136·4p
6	1·32	48	10·56	31p	141·0p
8	1·76	50	11·00	32p	145·5p
10	2·20	52	11·44	33p	150·0p
12	2·64	54	11·88	34p	154·6p
14	3·08	56	12·32	35p	159·1p
16	3·52	58	12·76	36p	163·7p
18	3·96	60	13·20	37p	168·2p
20	4·40	62	13·64	38p	172·8p
22	4·84	64	14·08	39p	177·3p
24	5·28	66	14·52	40p	181·9p
26	5·72	68	14·96	41p	186·4p
28	6·16	70	15·40	42p	191·0p
30	6·60	75	16·50	43p	195·5p
32	7·04	80	17·60	44p	200·0p
34	7·48	85	18·70	45p	204·6p
36	7·92	90	19·80	46p	209·1p
38	8·36	95	20·90	47p	213·7p
40	8·80	100	22·00	48p	218·2p

This chart is used by motorists to convert litres to gallons
and to compare pump prices per litre and per gallon.
When petrol is 39p per litre this is equivalent to 172·8p per gallon.

8. How many gallons would be equal to:
   (a) 7 litres?
   (b) 23 litres?
   (c) 83 litres?

9. Find the cost when petrol is
   39p per litre of:
   (a) 10 litres
   (b) 5 gallons
   (c) 15 gallons.

10. The petrol tank of a small car holds just over 7 gallons when full.
    (a) How many litres is this equivalent to?
    (b) How much would it cost to completely fill the tank with petrol
        at 38p per litre?

12 12 and 24 Hour Clock Times

12 Hour Times

12 hour clock times are used for most everyday practical purposes. Morning times are written with 'a.m.' afterwards i.e. 8.15 a.m., 10.30 a.m. etc. Afternoon and evening times are written with 'p.m.' afterwards i.e. 7.45 p.m., 11.25 p.m. etc. ...

T.V. viewing times are usually given in 12 hour time.

Exercise 12.1

Here is the BBC1 schedule for a Saturday evening.

1. (a) How long does Dynasty last?
   (b) How long does the Martian Chronicles last?
   (c) How much time is devoted to news, sport and weather?
   (d) How much longer is the film 'Trouble in the High Timber Country' than Wogan?

BBC 1

5.10	News, weather
5.20	Sport/regional news
5.25	The Dukes of Hazzard: Nothin' but the Truth
6.15	Jim'll Fix It: Jimmy Savile realises some more viewers' dreams
6.50	Trouble in the High Timber Country: 1980 film starring Eddie Albert, James Sloyan and Belinda Montgomery (See Films)
8.25	Paul Squire Esq: starring impressionist Paul Squire, with Melba Moore and Little Foxes
9.00	News, sport and weather
9.15	Dynasty: Nick wants Krystle to divorce Blake, who is himself being threatened by a Las Vegas racketeer
10.05	Wogan: Terry Wogan with guests, music and entertainment
10.50	The Martian Chronicles: sci-fi epic starring Rock Hudson. Earth has been obliterated by nuclear war; the handful of settlers on Mars are on their own
12.25–12.30	Weather

24 Hour Times

The 24 hour clock is used mainly for bus, train and airline timetables. It records the time from midnight to midnight instead of showing 'a.m.' or 'p.m.' times.

24 hour times always use **four figures**. The dot or point separates the hours from the minutes.

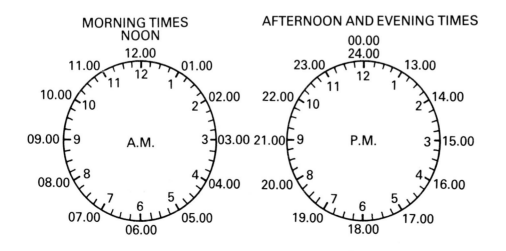

Examples

3 a.m.	is	03.00
6.37 a.m.	is	06.37
7 p.m.	is	19.00
10.47 p.m.	is	22.47

12 12 and 24 Hour Clock Times

2. Change to 24 hour time.
   (a) 6.52 a.m.
   (b) 3.46 p.m.
   (c) 7.59 p.m.
   (d) Quarter to midnight
   (e) Twenty to three in the morning.

3. Change to 12 hour times, remembering to write 'a.m.' or 'p.m.'
   (a) 07.45
   (b) 00.15
   (c) 23.45
   (d) 20.05
   (e) 21.17

4. This digital clock is 7 minutes slow. It should really show 18.05.

   Write out how many minutes 'fast' or 'slow' are the following clocks.

	TIME SHOWN	CORRECT TIME
(a)	14.26	14.17
(b)	01.11	01.15
(c)	23.59	00.02
(d)	17.01	16.53
(e)	13.49	14.02

5. How many hours between:
   (a) 6 a.m. and 4 p.m.?
   (b) 04.15 and 23.15?
   (c) 00.30 and 15.30?
   (d) 17.20 and 14.20 the following day?

6. How many minutes between:
   (a) 10.45 a.m. and 2.16 p.m.?
   (b) 12.47 and 14.03?
   (c) 19.58 and 22.36?
   (d) 21.47 and 00.17 the next morning?

7. (a) A train left London at 07.25 and arrived at Newcastle 5 hours 55 minutes later. What was its time of arrival?
   (b) A ship left Felixstowe at 11.34 on a Wednesday and arrived at Oslo 22 hours and 43 minutes later. What was its time of arrival on Thursday?

Time Differences

Denmark sticks to Central European time (GMT + 1 hour) as does most of the continent. Denmark's capital Copenhagen is 1 hour ahead of London time. New York time is 5 hours behind London time.

In summer the time difference in some major world cities looks like this:

Los Angeles	Chicago	New York	London	Sydney	Auckland	Copenhagen	Moscow
4 a.m.	6 a.m.	7 a.m.	NOON	9 p.m.	11 p.m.	1 p.m.	3 p.m.

Examples

London time (10.30) minus 8 hours = Los Angeles time (02.30)
London time (09.45) plus 11 hours = Auckland time (20.45)

8. When it is 13.45 in London, what time is it in:
   (a) Chicago?
   (b) Moscow?
   When it is 16.20 in Sydney, what time is it in:
   (a) Copenhagen?
   (b) Moscow?

9. When it is 08.15 in Moscow, what time will it be in:
   (a) Auckland?
   (b) New York?
   When it is quarter past midnight in Auckland, what time will it be in:
   (a) New York?
   (b) Los Angeles?

10. Flights from London to New York take $5\frac{1}{2}$ hours and flights from London to Sydney $22\frac{1}{2}$ hours. Give the times of arrival in **local** time for the following flights:
    (a) 10.15 London to New York
    (b) 08.15 London to Sydney
    (c) 23.30 London to Sydney

12 12 and 24 Hour Clock Times

Time-tables

Shown below is the time-table for a bus route.

SMALLBOROUGH—BIGTOWN via Middleton

	Mondays to Fridays							Saturdays					
Smallborough	0714	0758	0933	1213	1506	1632	1747	0753	0843	0933	1213	1353	1603
Rose & Crown	0720	0804	0939	1219	1512	1638	1753	0759	0849	0939	1219	1359	1609
Middleton	0734	0818	0953	1233	1526	1652	1807	0813	0903	0953	1233	1413	1623
Bigtown North	0746	0830	1005	1245	1538	1704	1819	......	......	1005	1245	1425	1635
Bigtown Stn	......	0840	1015	1255	1548	......	......	......	......	1015	1255	1435	1645

BIGTOWN—SMALLBOROUGH via Middleton

	Mondays to Fridays							Saturdays					
Bigtown Stn	......	0850	1020	1300	1600	......	......	......	......	1020	1300	1440	1650
Bigtown North	0752	0900	1030	1310	1610	1710	1825	......	......	1030	1310	1450	1700
Middleton	0804	0912	1042	1322	1622	1722	1837	0818	0908	1042	1322	1502	1712
Rose & Crown	0818	0926	1056	1336	1636	1736	1851	0832	0922	1056	1336	1516	1726
Smallborough	0824	0932	1102	1342	1642	1742	1857	0838	0928	1102	1342	1522	1732

NO SERVICE SUNDAYS OR PUBLIC HOLIDAYS

Exercise 12.2
Refer to the bus time-table.

1. How many minutes does it take to travel from Smallborough to Bigtown Station?

2. If you want to travel from Smallborough to Bigtown Station on a Saturday
   (a) At what time is the first bus you can catch?
   (b) At what time will you arrive?

3. How many minutes does it take to travel from Bigtown Station to Rose and Crown?

4. How many minutes does it take to travel from Middleton to Bigtown North?

5. If you arrive at Bigtown Station at 1.30 p.m. on a Saturday afternoon
   (a) At what time is the next bus to Smallborough?
   (b) For how long must you wait?

6. How long does it take to travel from Middleton to Smallborough?

7. If the 12.13 from Smallborough is delayed for 9 minutes in a traffic jam at what time will it arrive at Bigtown North?

8. If you catch the 13.53 from Smallborough on a Saturday, how many hours and minutes could you spend at Bigtown Station before catching the last bus back to Smallborough?

9. If the last bus from Bigtown North to Smallborough on a week day arrives 23 minutes late, at what time would it arrive at
   (a) Rose and Crown?
   (b) Smallborough?

10. If you reach Middleton at 1.55 p.m. on a week day
    (a) By how many minutes have you missed your bus to Smallborough?
    (b) How long must you wait for the next one?

13 Ratio and Proportion

To make concrete, ballast (sand and gravel) is mixed with cement and water.

To make the mixture correctly the ballast and cement must be mixed in the **right proportion**. The amounts of ballast and cement should be in the ratio 5:1 (the ratio sign : means 5 parts of ballast for 1 part of cement).

Example
What quantities of ballast and cement wuld be required to make up 1200 kg of concrete? The ballast and cement are mixed in the ratio of 5:1.
Add the ratio parts together 5 + 1 = 6
6 parts, so each part is 1200 ÷ 6 = 200 kg
Ballast is 5 parts = 200 × 5 = <u>1000 kg</u>
Cement is 1 part = 200 × 1 = 200 kg

Exercise 13.1

1. To make a short pastry, fat and flour is mixed in the ratio 1:2. How much fat is used to make 600 g of pastry?

2. A man dies and leaves £2400 to be divided between his wife and his daughter in the ratio 5:7. How much does each receive?

3. An alloy consists of two metals A and B in the ratio 5:6. What weight of each metal if the total weight of the alloy is 55 kg?

4. A firm employs 500 people. The ratio of males to females is 7:3. How many females does the firm employ?

5. Three directors of a company hold shares in the ratio 1:2:3. What share should each receive if the company makes a profit of £21000?

Sometimes we have to find One of the Parts

Example   A sum of money is divided between Bob and Harry in the ratio 5:4.
(a) How much would Bob receive if Harry received £240?
(b) What was the sum of money?

(a) Harry received 4 parts which is £240.
So 1 part is $\frac{£240}{4}$ = £60
So, Bob receives 5 parts which must be
£60 × 5 = £300

(b) The Sum
= Bob's share + Harry's share
=   £300   +   £240   =   £540

6. The ratio of Jane's age to Elizabeth's age is 5:4. Find Elizabeth's age if Jane is 25 years old.

7. A sum of money is divided in the ratio 3:7. Find the largest part if the smallest part is £15.

8. An alloy is made up by weight of 13 parts tin and 12 parts lead. If the tin weighs 390 grams, what must be the weight of the lead?

9. A sum of money is divided between A, B and C in the ratio 3:4:5.
(a) Find out how much A and B received if C received £30.
(b) Find the sum of money.

10. To make up a stew a canteen uses meat, potatoes and other vegetables in the ratio 2:3:4. How much meat and other vegetables are used if the weight of potatoes is 15 kg?

13 Ratio and Proportion

Direct Proportion

£100 = 357 marks
So, £200 = 714 marks

The **more** pounds you change
the **more** marks you receive

and, £10 = 35·7 marks

The **fewer** you change
the **less** you receive.

The amount you receive is **directly proportional** to the amount you change.

When carrying out calculations in direct proportion we find the cost or the weight of **one**.

Example   5 Similar books cost £12·50. How much for 12 books?

   5 cost £12·50
So,   1 costs £12·50 ÷ 5 = £2·50

and,   **12 will cost** £2·50 × 12 = £30

Exercise 13.2

1. 20 metres of cable cost £4. How much for 75 metres?

2. A man earns £273 in 3 weeks. How much will he earn in 9 weeks?

3. A car travels 105 miles on 3 gallons of petrol. How far will it travel on 7 gallons?

4. A lady earns £95 for working 38 hours.
   (a) How much does she earn for working 10 hours?
   (b) How many hours must she work to earn £40?

Use your calculator for the following:

5. 597 Spanish pesetas can be exchanged for £3.
   (a) How many pesetas for £11?
   (b) How many pounds for 3980 pesetas?

6. A 17 day holiday in Portugal costs £368·05. How much would it cost for 11 days at the same rate?

7. 10·3 kg of lamb chops cost £24·72.
   (a) How much for 3·9 kg?
   (b) How many kg could I buy for £15·60?

8. 11·70 USA dollars can be exchanged for £7·80.
   (a) How many dollars for £9·25?
   (b) Find to the nearest penny, how much for 11·25 USA dollars.

9. 1254·6 Greek drachmas can be exchanged for £10·20.
   (a) How many drachmas for £3·70?
   (b) Find to the nearest penny, how much for 10000 drachmas.

10. A small factory uses 733·4 units of electricity to run its machinery during a 38 hour working week.
    (a) How many units of electricity will it use in 9 hours?
    (b) If the factoy is charged 4·19p per unit, find to the nearest penny, the hourly cost of electricity.

14 Graphs and Charts

Sometimes information can be conveyed more easily by pictures or diagrams.

In a pictograph symbols are used to represent information. The pictograph shown opposite shows how a firm's workforce has increased between 1940 and 1980. One 'matchstalk man' represents 100 workers.

In 1940 the firm employed 150 workers.

Exercise 14.1

1. How many workers did the firm employ in
   (a) 1960
   (b) 1970
   (c) Which 10 year period saw the biggest increase in the size of the workforce?

2. The pictograph opposite shows how a small oilfield's production has increased over a 5 year period. One barrel is drawn to represent 10000 litres. In 1977 the oilfield produced 15000 litres.

   How many litres were produced in
   (a) 1979   (b) 1981

   (c) Find the total amount produced in the 5 year period.
   (d) Find the average production per year over this period.

3. This bar chart tells us how many **new** cars were sold by a certain garage between 1976 and 1979. All graphs and charts have scales. In this 1 mm on the vertical scale represents 1 car (10 cars = 1 cm) and the columns are drawn 10 mm (1 cm) thick.
   How many cars were sold in:
   (a) 1977?
   (b) 1979?
   (c) What has the average number of cars sold per year over the four year period?
   (d) Which period saw the biggest increase in new car sales?

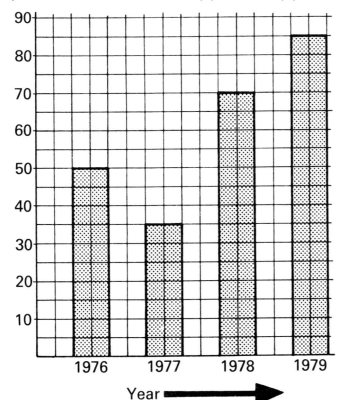

39

14 Graphs and Charts

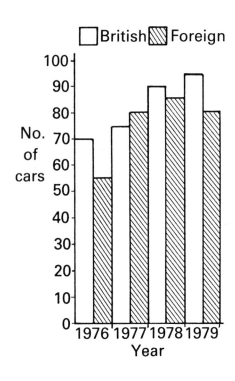

4. This **double** bar chart compares sales of **used** British and foreign cars over the same period at the same garage. It uses the same scale.

   How many foreign cars were sold in
   (a) 1976?   (b) 1978?
   (c) What was the total number of used cars sold in 1979?
   (d) In which years did the sales of foreign used cars exceed the sales of British used cars?

5. Use squared paper to draw a bar chart showing a shop's daily takings. Use a vertical scale of 10 mm (1 cm) = £100, make the bars 10 mm thick and space them 10 mm apart.

Day	Mon	Tues	Weds	Thurs	Fri	Sat
Takings	£350	£450	£500	£200	£650	£950

   (a) Which is the probable 'early closing' day?
   (b) Find the weeks total takings.
   (c) Find the average daily takings.

This pie-chart shows the result of a survey carried out before a by-election.
A number of people were questioned as to how they would vote.

In a pie-chart each quantity is represented by an angle and by a sector of a circle.
The angle represented by those who said they would vote Labour is:

$$\frac{30}{100} \times \frac{360}{1} = 108°$$

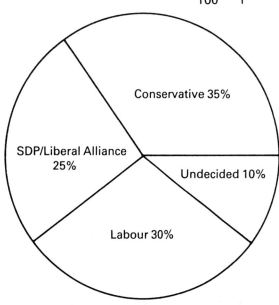

6. In the pie chart what angle is represented by:
   (a) Conservative
   (b) SDP
   If 160 people were questioned for the survey how many:
   (c) Said they would vote Labour?
   (d) Were undecided?

7. Draw a pie-chart to show the yearly exports of a firm making home computers. The percentage sales to various countries are as follows. Germany 30%, Italy 25%, France 20%, USA 10%, Japan 5%, Others 10%. Use a radius of 50 mm for the pie-chart. If in 1983 the firm exported 72000 units how many were sold in
   (a) Japan
   (b) Germany
   (c) USA
   (d) By how many units did sales to Italy exceed sales to France?

14 Graphs and Charts

8. In a straight line graph, points are marked to represent data and then joined up by a line. This graph can be used to convert miles to kilometres and vice versa from the original information that 8 km = 5 miles and 40 km = 25 miles.

On the vertical scale 1 mm = 1 mile and on the horizontal scale 1 mm = 1 km.
From the graph we can find that 20 miles = 32 km.
   (a) How many miles are represented by 24 km?
   (b) How many km are represented by 30 miles?
   (c) How many miles are represented by 90 km?
   (d) How many km are represented by 45 miles?

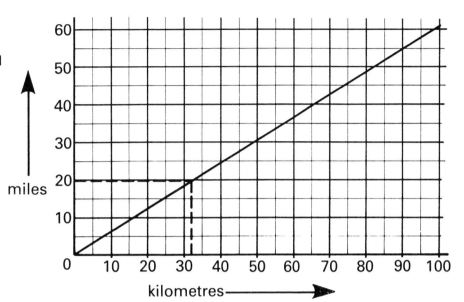

9. Use squared paper to draw a straight line graph which converts pounds (lbs) to kilograms and vice versa. Use a vertical scale of 10 mm = 10 kg and a horizontal scale of 10 mm = 10 lbs. Plot two points using the information 10 kg = 22 lbs and 50 kg = 110 lbs, draw the straight line graph and use it to convert:
   (a) 20 kg to lbs.
   (b) 40 lbs to kg.
   (c) 40 kg to lbs.
   (d) 100 lbs to kg.

10. The graph below shows how a hospital patients body temperature varied over a 6 day period. 37° represents normal temperature, 38° feverish and 39°–40° represents high fever. The temperatures were taken twice each day.

   What was the patients body temperature on
   (a) Saturday p.m.?
   (b) Wednesday a.m.?
   (c) At what time did the patients temperature reach its maximum?
   (d) Between which 24 hour period did the temperature show the 'largest' decrease?

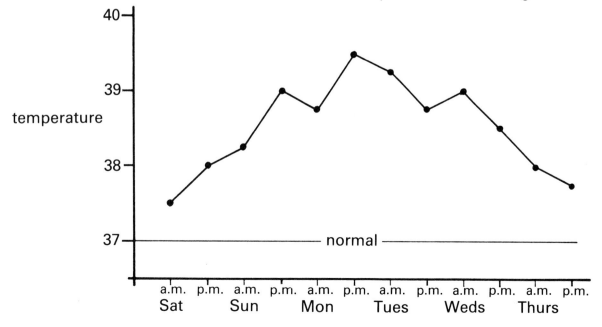

41

15 Areas and Perimeters

Area is the measure of the amount of surface a shape covers and is measured in square units, i.e. m^2, cm^2, ft^2, in^2 etc. . . .

The perimeter is the distance around a shape and is measured in units i.e. m, cm, ft, ins etc. . . .

Example

This rectangle has a length of 3·5 metres and a width of 2 metres.

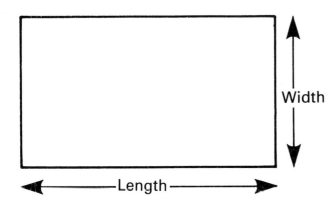

The Area = length × width
$$A = l \times w,$$
$$= 3·5 \times 2,$$
$$= 7 \, m^2$$

The Perimeter = (2 × length) + (2 × width)
$$= 2l + 2w,$$
$$= (2 \times 3·5) + (2 \times 2),$$
$$= 7 + 4,$$
$$= 11 \, m$$

Exercise 15.1

1. A rectangle has a length of 5 cm and a width of 4·2 cm. Find:
   (a) Its area.
   (b) Its perimeter.

2. This rectangle has an area of 14 m^2 and a length of 4 m.

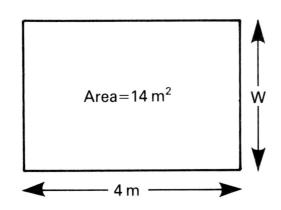

Find:
(a) Its width
(b) Its perimeter.

3. This rectangle has a perimeter of 18 cm and a length of 5 cm. Find:
   (a) Its width   (b) Its area.

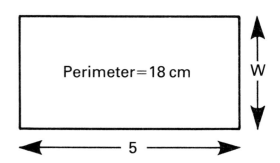

4. The area of a triangle is formed by using the formula:
   Area = $\frac{1}{2}$ × base × height

This triangle has a base of 6 cm and a height of 4·5 cm.

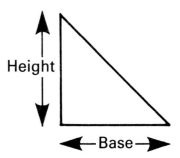

$A = \frac{1}{2} \times b \times h$
$= \frac{1}{2} \times 6 \times 4·5$
$= 13·5 \, cm^2$

Find the areas of the following triangles:
(a) Base 3·8 metres, height 4 metres.
(b) Base 16 metres, height 2·9 metres.

15 Areas and Perimeters

5. (a) This triangle has an area of 52 mm² and a base of 13 mm. Find its height.

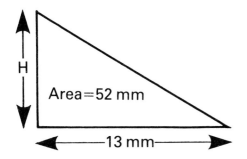

(b) This triangle has an area of 24 m² and a height of 6 m. Find its base.

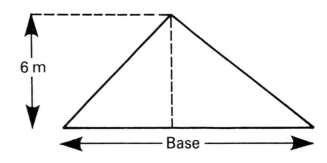

6. The area of a circle is found by using the formula

Area = $\pi \times$ radius $\times$ radius
$A = \pi r^2$

π (PI) is the ratio of the circumference to the diameter and always has a value slightly more than 3. For calculations we can use $\pi = 3\cdot 1$.

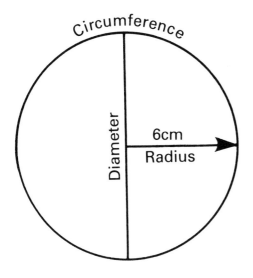

This circle has a radius of 6 cm.
$A = 3\cdot 1 \times 6 \times 6$
$ = 111\cdot 6$ cm²

Find the areas of circles with
(a) radius 9 cm
(b) radius 2·5 m (use a calculator)
(c) diameter 11·5 cm (use a calculator)

7. The distance around a circular shape is called the **circumference**. The circumference is found by using the formula

Circumference = $2 \times \pi \times$ radius
$C = 2\pi r$

This circle has a radius of 7 cm.
$C = 2 \times 3\cdot 1 \times 7$
$ = 43\cdot 4$ cm

Use a calculator to find the circumferences of circles with:
(a) radius 1·7 m
(b) diameter 5·6 cm
(c) radius 11·9 cm.

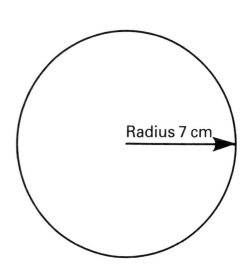

15 Areas and Perimeters

8. Find the areas of the shapes shown below.

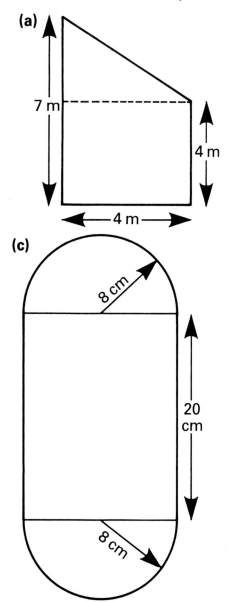

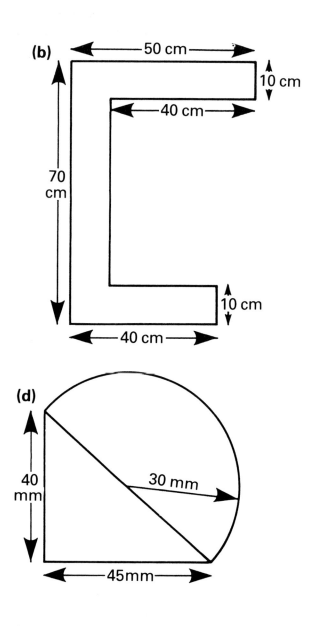

9. Find the area of the two washers.

10. A bathroom area measuring 2·1 m by 1·2 m is to be tiled. The tiles measure 15 cm by 15 cm and are sold in boxes of 12. The cost of 1 box is £3·75. Find:

   (a) The number of tiles required.
   (b) The number of boxes.
   (c) The total cost.

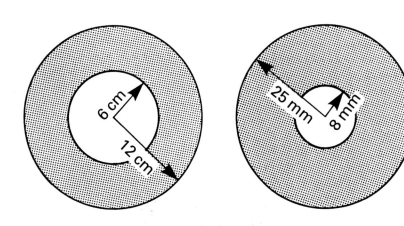

16 Volume

Volume is a measure of the amount of space a shape occupies and is measured in cubed units i.e. m³, cm³, ft³, in³ etc. . . .

Example

This cuboid has a length of 6 cm, a width of 5 cm and a depth of 8 cm.

Volume = length × width × depth.
$$V = l \times w \times d$$
$$= 6 \times 5 \times 8$$
$$= 240 \text{ cm}^3$$

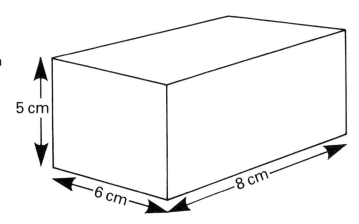

Exercise 16.1

1. Find the volumes of the shapes shown below:

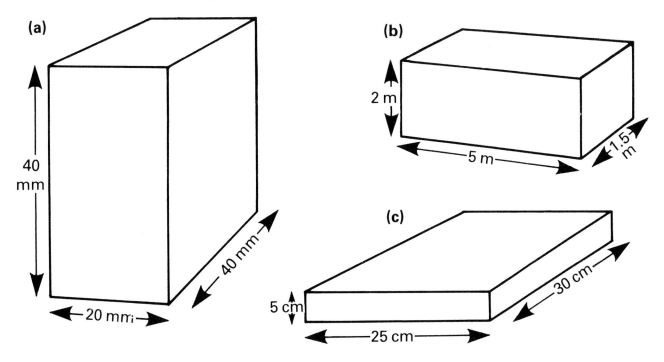

2. The volume of a prism is found by multiplying the area of the triangular face by the depth.

$$V = \tfrac{1}{2} \times b \times h \times d$$
$$= \tfrac{1}{2} \times 8 \times 4 \times 12$$
$$= 192 \text{ cm}^3$$

Find the volume of a prism with base 12 cm, height 10 cm and depth 15 cm.

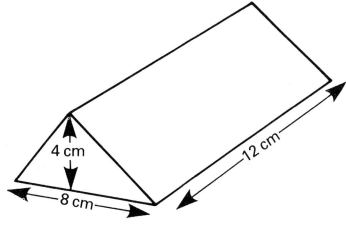

16 Volume

3. Find the volumes of the shapes shown below

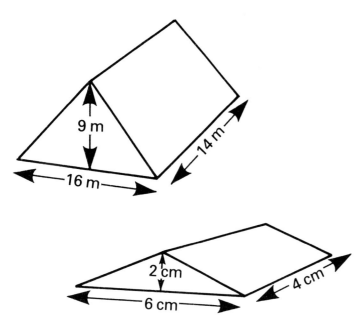

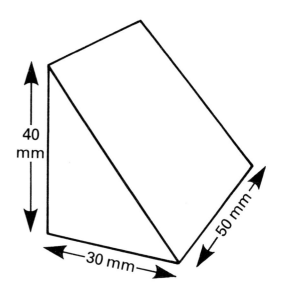

4. The volume of a cylinder is found by multiplying the area of the circular face by the depth.

$V = \pi r^2 d$
$v = 3 \cdot 1 \times 10 \times 10 \times 15$
$ = 4650 \text{ cm}^3$

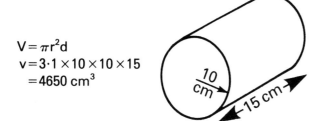

Find the volume of a cylinder with a radius of 4 metres and a depth of 2 metres.

5. Find the volumes of the shapes shown below

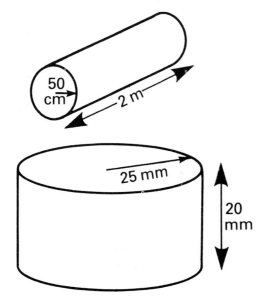

6. This cuboid has a length of 20 cm, a width of 12 cm and a volume of 1920 cm². Find its depth.

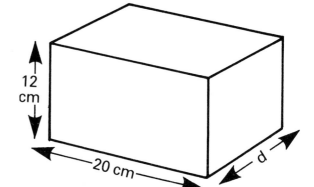

7. How many boxes of chalk measuring 6 cm by 2 cm by 9 cm could be fitted in a larger box measuring 60 cm by 4 cm by 9 cm.

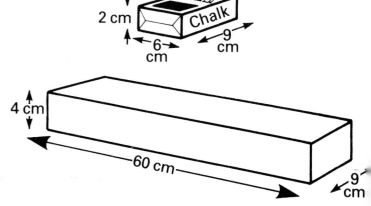

16 Volume

8. What volume will remain after a hole of diameter 40 mm has been bored through the cuboid shown? ($\pi = 3\cdot1$)

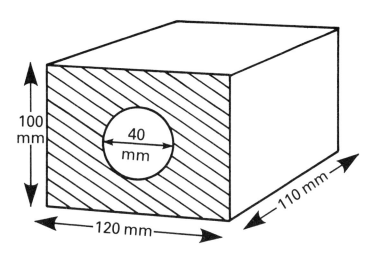

9. Find the volumes of the shapes shown below.

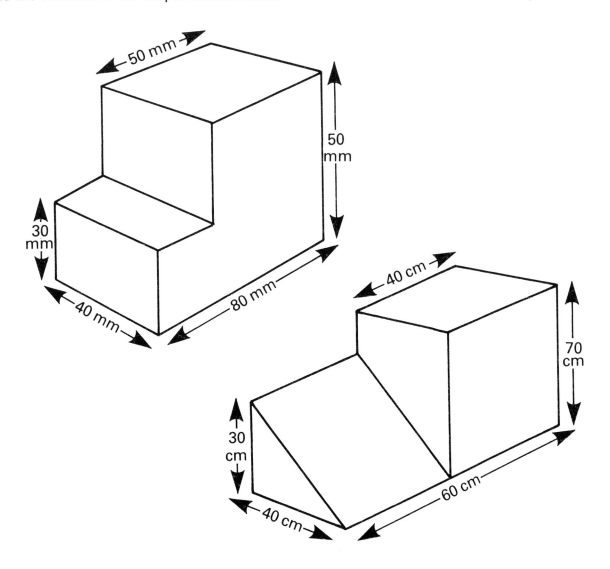

10. An area measuring 3·5 metres by 5 metres is to be filled with concrete to a depth of 200 mm. What volume of concrete will be needed?

17 Scales and Maps

Scale drawings are used by people who design, build and invent.

Shown below is a scale plan of a bungalow. The scale used is 1 cm to 1 m.

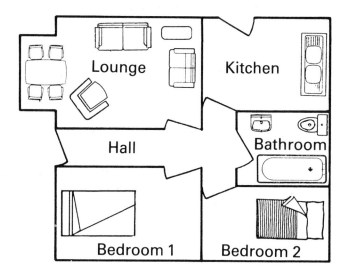

The scale length of bedroom 1 is 4 cm. This means the **real** (or actual) length is 4 metres.

Exercise 17.1

1. Copy out the table below, use a centimetre ruler to measure up the rooms.

	SCALE		REAL	
	LENGTH	WIDTH	LENGTH	WIDTH
KITCHEN				
BEDROOM 1	4 cm		4 m	
BEDROOM 2				
BATHROOM				

2. What is the scale area of:
   (a) The kitchen?
   (b) Bedroom 2?
   What is the real area of:
   (c) The bathroom?
   (d) Bedroom 1?

3. Find by measuring:
   (a) The **scale** area of the lounge.
   (b) The **real** area occupied by the bungalow.

4. The owner of the bungalow decides to build a garage at the side.
   A builder submits an estimate equivalent to £142 for each square metre of ground the garage will occupy. The owner decides to build the garage himself and finds the cost to be £70 for each square metre. Find how much the owner saved by building the garage himself.

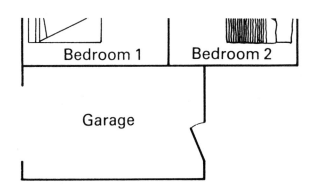

5. A scale model of a car has a length of 10 cm. In reality the car is 5 metres long. Copy out and complete the table below which compares the real and scale sizes of the car.

	SCALE MODEL	REAL MODEL
LENGTH	10 cm	5 metres
HEIGHT		1.5 metres
WIDTH	2.5 cm	
WHEEL		50 cm

17. Scales and Maps

6. This scale map shows approximate distances by road between five towns in Yorkshire. Measure the distances to find how far it is by road from:
   (a) York to Leeds.
   (b) Leeds to Castleford.
   (c) Goole to York via Selby.
   (d) Leeds to Goole via Castleford and Selby.

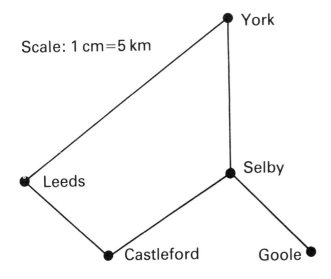

Scale: 1 cm = 5 km

7. A man travels by car each weekday from Goole to his work in York.
   (a) Use the map to find the distance for his 'round trip'.
   (b) How far would he travel in one week if he only used his car for work?
   If the petrol consumption of his car is 14 kilometres per litre:
   (c) How many litres would he use in that week?
   If the cost of petrol is 37p per litre:
   (d) How much would he have to spend on petrol?

8. The distance from London to Rome on the map below when measured with a millimetre rule is 80 mm. This means that the real flying distance is 800 miles. Use a millimetre ruler to find the **real** flying distances from:
   (a) London to Frankfurt.
   (b) London to Alicante.
   (c) Paris to Barcelona.

9. A private aircraft leaves London for Malaga landing at Marseilles and Palma on the way. What was the total distance the plane travelled?

10. Find the average speed of aircraft flying the following routes and times:
    (a) London to Palma in 2 hours.
    (b) Frankfurt to Paris in ½ hour.
    (c) Paris to Madrid in 3 hours.

─500─ distances from London

500 nautical miles is approximately one hour in air cruising time

This aeroplane route map shows distances from London to various cities in Europe. The scale used is 1 mm = 10 nautical miles.

18 The City Hall

This city hall is used for many different events; exhibitions, dances plays etc.
For plays, concerts and musicals, seats have to be placed in the hall.
Shown below is part of the seating plan.
Downstairs there are 26 rows, each row having the same number of seats. These rows are lettered A–Z. Upstairs there are 15 rows, each having the same number of seats. These rows are lettered AA–OO. Use the plan to answer the questions in Exercise 18.1.

Exercise 18.1

1. (a) How many seats are downstairs?
   (b) How many seats are upstairs?
   (c) What is the total seating capacity of the town hall?

2. For a play tickets are priced as follows:
   Downstairs: Rows A–H £4·50
              Other seats £4·00
   Upstairs: Rows AA–DD £5·00
         Other seats   £4·00
   (a) How much for 3 seats in row E?
   (b) How much for a ticket marked KK15?
   (c) How much would I pay in all for tickets marked CC17, CC18, P20 and P21?
   (d) What would be the total takings if all the tickets were sold?

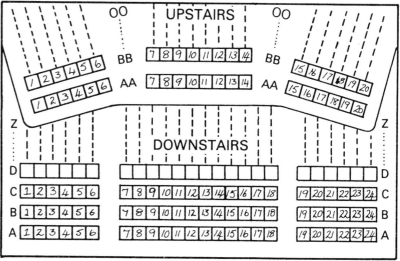

3. For another play all the seats were at the same price. 75% of the tickets for downstairs and 80% of the tickets for upstairs were sold.
   (a) What was the total number of tickets sold?
   (b) How many upstairs tickets were **unsold**?
   (c) If the receipts were £2124, what was the price per ticket?

4. A rock group hire the city hall for a concert. The charges are:
       £6·50 downstairs
       £5·00 upstairs
   They manage to sell 550 downstairs tickets and 75% of the upstairs tickets.
   (a) How many tickets were **unsold**?
   (b) If the hire of the hall and other expenses amounted to £1200, how much profit did the group make?

5. For a musical the stage has to be enlarged and this means that the first six rows of seats have to be removed.
   (a) What is the new seating capacity of the city hall?
   (b) On the opening night all the downstairs seats were priced at £5 and all the upstairs seats at £6. How much was taken if the hall was completely full?
   (c) At a matinee the downstairs was filled to 75% of its capacity and upstairs to 80% of its capacity. How many seats were empty?

19 The Garden

Exercise 19.1

Opposite is the plan of a garden.

The scale is 1 cm to 2 metres.

1. (a) What is the real length?
   (b) What is the scale area?
   (c) What is the real area?

2. The owners plan to set out the garden as shown opposite.

   What are the real areas of:
   (a) The Patio.
   (b) The Lawn.
   (c) The Vegetable section.
   (d) The Path.
   (e) The border around the lawn.

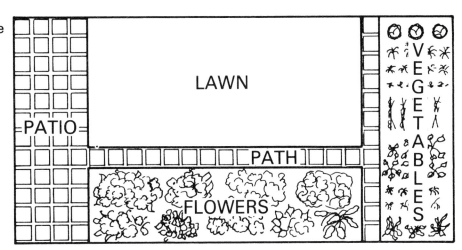

3. **The Patio and the Path—** These are formed by laying ballast to a depth of 15 cm and mortaring paving stones to the ballast. Each paving stone measures 50 cm by 50 cm.
   (a) How many cubic metres of ballast would be needed?
   (b) How many paving stones would have to be purchased?
   (c) What would be the cost of purchasing the ballast and the paving stones? The paving stones are £1·50 each and the ballast is sold at £12 per cubic metre. (The ballast must be purchased in whole cubic metres).

4. **The Lawn—**This is to be made with turf. Strips of turf 1 metre wide are to be laid lengthways along the lawn area.
   (a) How many strips of turf will be required?
   (b) What would be the total cost if one square metre is 90p?

5. **The Flower section—** Flowering shrubs are to be planted at the rate of three for every square metre. The average price for the shrubs is 37p each.
   (a) Find the number of flowering shrubs to be planted.
   (b) Find the total cost of buying the shrubs.

20 The Rag Trade

Exercise 20.1

This diagram represents a scale drawing of a template used in the manufacture of circular skirts.

Two pieces need to be cut to make one size 14 skirt.

1. Use your calculator to find:
   (a) The area of material in one piece, and
   (b) The area of material used in making one skirt.
   (c) How much material will be wasted in one cut if the material used measures 84 cm by 84 cm?
   (Take $\pi = 3.1$)

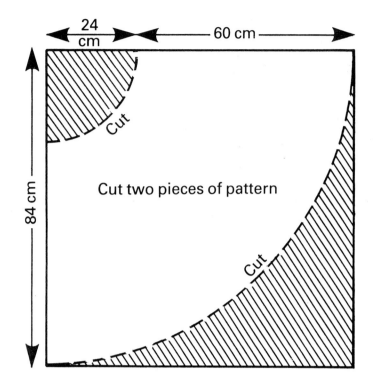

2. The textile factory who manufacture the skirts pay their workers £1·40 per hour plus 35p for each skirt they make.

   Calculate the gross weekly wage for:
   (a) Liz Patterson who works 40 hours and makes 48 skirts.
   (b) Jenny Spencer who works 37 hours and makes 45 skirts.
   (c) Shirley Winter who works 29 hours and makes 37 skirts.

3. The factory reckon that the average time taken by a worker to make a skirt is 40 minutes. Use this figure to calculate how many skirts should be made by:

   (a) One worker in 32 hours.
   (b) Two workers in 36 hours.
   (c) Four workers in 38 hours.

The overall cost of manufacturing each skirt is £5·50.

4. If the factory sells the skirts at £6·05, calculate:
   (a) The percentage profit on one skirt.
   (b) The total profit on a weeks production of 4800 skirts.

5. A wholesaler offers to buy 10000 skirts from the factory on condition that the selling price of £6·05 per skirt is reduced by 22p. Calculate:
   (a) The total profit the factory will make if they agree.
   (b) The new profit percentage.

21 The Bakery

Exercise 21.1

This recipe is used by a small bakery and makes approximately ½ kg of rising dough—enough to make one standard ½ kg white loaf.

1. Express in their lowest terms the ratios of
   (a) flour to margarine,
   (b) flour to year,
   (c) margarine to yeast.

2. The bakery receives an order for 500 standard white loaves. Use the recipe to find the weights in kilograms of the following ingredients that they must use:
   (a) flour   (b) margarine   (c) fresh yeast.

3. The chart below shows the number of standard white loaves the bakery sold in a certain week. Copy out the chart in your book, complete it and show the weekly totals.

DAY	NUMBER OF LOAVES	FLOUR IN kg	MARGARINE IN kg	FRESH YEAST IN kg
Mon	400	120	10	6
Tues	300		7·5	4·5
Wed	300			
Thurs	350	105	8·75	5·25
Fri	600			
Totals	1950			

21 The Bakery

4. Each year the bakery takes orders for Christmas cakes made from their own traditional recipe.

Ingredients

300 g currants
300 g sultanas
200 g raisins
75 g chopped almonds
100 g cherries
100 g chopped peel
250 g plain flour
250 g butter
250 g sugar
5 eggs
3 tablespoons of brandy

Last Christmas 240 of these cakes were sold. What weight in kilograms of each of the following were used?
(a) Currents
(b) raisins
(c) flour
(d) cherries
(e) How many **dozen** eggs were used?
(f) What is the ratio of butter to chopped almonds?

5. Recipes in cookery books usually give instructions as regards cooking times and temperatures to use. The chart below can be used to convert gas cooker temperatures to Celcius and Fahrenheit electric cooker temperatures. For example, GAS MARK 1 = 275°F = 130°C.

Use the chart to answer the following

(a) What gas mark is equivalent to 175°C?
(b) What Fahrenheit temperature is gas mark 3 equivalent to?
(c) What Fahrenheit temperature is the same as 220°C?
(d) What gas mark is equivalent to 400°F?

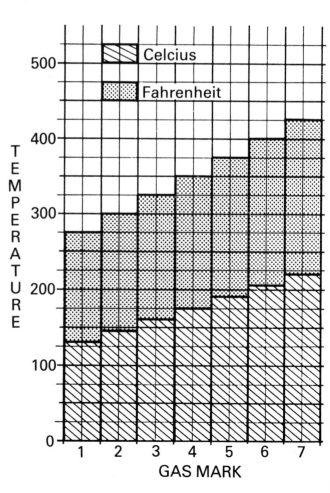

22 Angles, Gradients and Bearings

Angles

An angle measures the amount of turn or steepness.

One complete turn is divided into 360 equal parts and each part is called a degree. [One complete turn = degrees (360°)]

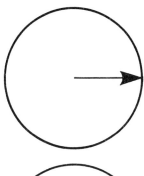

One half turn = 180° (A straight line)

One quarter turn = 90°
Commonly known as a **right angle**. The lines are said to be **perpendicular** to each other.

An instrument for measuring angles is a **protractor**

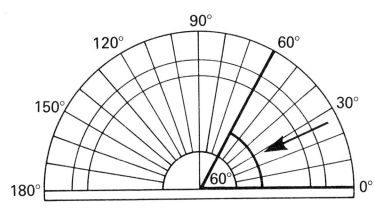

This angle measures 60°

Use a protractor to measure the following angles.

1.
2.
3.
4.
5.

Exercise 22.1

Use a protractor, ruler and pencil to construct angles of

1. 55°
2. 65°
3. 110°
4. 170°
5. 95°

22 Angles, Gradients and Bearings

Gradients

Sometimes the amount of slope or steepness of a road or incline is referred to as the gradient.

This road sign is a warning to drivers that they are approaching a steep hill with a 1 in 4 gradient or climb.

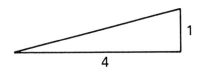

This road sign warns drivers that they are approaching a hill with a downward slope of 1 in 6.

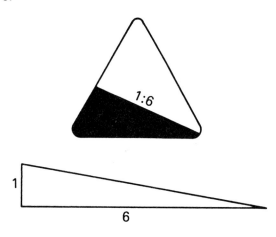

A would be steeper than B as 1:4 ($\frac{1}{4}$) is a greater fraction than 1:6 ($\frac{1}{6}$)

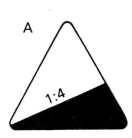 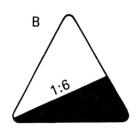

Exercise 22.2

1. Which would be the steepest hill?

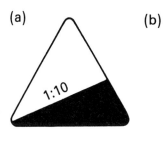 

2. Use squared paper to draw the following gradients.
   (a) 1:3   (b) 1:8   (c) 1:6   (d) 1:10

3. More modern warning signs give the gradient as a percentage.

   This sign indicates a hill with a downhill slope of 10%

   i.e. $10\% = \frac{10}{100} = \frac{1}{10}$ or a slope of 1 in 10

   (a) Which would be the steepest hill?

   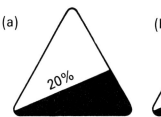 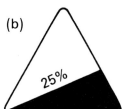

   (b) Use squared paper to draw the gradient of a hill with a 20% downward slope.

22 Angles, Gradients and Bearings

Bearings

A bearing is a way of giving the direction of one place from another using angles.

The angle used in giving bearings is usually measured in a **clockwise** direction from **North**.

Bearings are always given as three figure numbers.

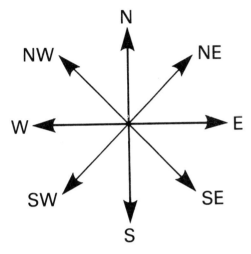

Examples

1. The bearing of X from Y is 075°

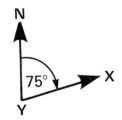

2. The bearing of Y from O is 140°

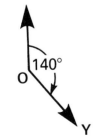

3. The bearing of Z from O is 215°

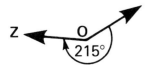

Exercise 22.3

1. Find the bearing of A from O in each diagram.

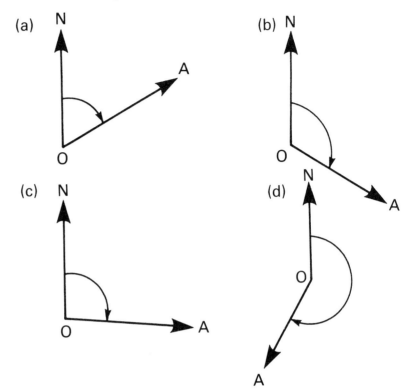

2. Here are four towns.
   Find the bearings of
   (a) Newtown from Penton
   (b) Routledge from Penton
   (c) Newton from Sutton
   (d) Routledge from Sutton.

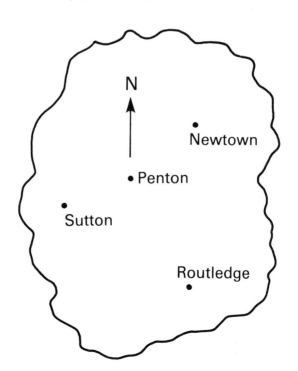

MONEY..........................→

23 Wages and Salaries

People who work can be paid in several different ways.

Wage earners are usually paid weekly. The amount they receive is usually based on the number of hours they work.
Other wage earners known as 'piece workers' receive an hourly rate plus a bonus for the amount of goods they produce.
Salesmen often receive a basic wage plus **commission** on the amount of goods they sell.

Salary earners, such as teachers and civil servants, receive a fixed amount each year and are ususally paid monthly.

The table below shows an engineering firm's rates of pay.

SMALLWORKS ENGINEERING LTD	
MECHANICS	£2·80 per hour
SEMI-SKILLED	£2·10 per hour
CLERICAL STAFF	£1·90 per hour
TRAINEES	£1·40 per hour

Rates are based on a 40 hour working week. Overtime at $1\frac{1}{2}$ times the normal rate Mondays to Saturdays and at 2 times the rate (double time) on Sundays.

Example

What would be a mechanic's wage for a week in which in addition to his normal 40 hours he worked 8 hours overtime between Monday and Saturday plus 4 hours overtime on Sunday?

Basic wage   = £2·80 × 40   = 112·00
Overtime at ×$1\frac{1}{2}$ = £2·80 × $1\frac{1}{2}$ × 8 = 33·60
Overtime at ×2   = £2·80 × 2 × 4 = 22·40
Total (or gross) wage   = 168·00

Exercise 23.1

1. Find the basic wages for the following workers.
   (a) Semi-skilled.
   (b) Trainee.

2. How much extra would the following workers be paid for the following overtime?
   (a) A semi-skilled worker, 8 hours at time and a half ($1\frac{1}{2}$).
   (b) A clerical worker for 4 hours on a Sunday (double time).

3. What would be the total (or gross) weekly wage of a semi-skilled worker if in addition to his normal working week he worked 4 hours at 'time and a half' plus 3 hours at 'double time'?

4. Here is a clock card for Pete Smith, a mechanic. Employees are allowed 2 minutes clocking in time.

NAME: P. Smith			WORKS NO: 1401			
	MORNING		AFTERNOON		HOURS	
	IN	OUT	IN	OUT	NORMAL	O/T
MON	07.59	12.01	13.00	17.00	8	
TUES	08.01	12.00	12.59	19.59	8	2
WED	08.00	11.59	13.01	17.01	8	
THURS	07.58	12.02	13.00	18.00	8	1
FRI	08.02	12.00	12.58	17.00	8	
SAT	08.01	11.00				3
SUN	09.00	12.00				3
TOTAL HOURS					40	9

Normal practice is to work an 8 hour day Monday to Friday with lunch from 12 noon to 1 p.m.
(a) How many hours at 'time and a half' did he work?
(b) How many hours at 'double time' did he work?
(c) Calculate his **gross** wage for the week.

23 Wages and Salaries

5. Shown below are Kate Brown's clock card and incomplete payslip for the same week.

(a) Use the amount shown under 'Basic Wage' to find out what sort of work she does.

The amount Kate actually receives NET WAGE = GROSS WAGE − TOTAL DEDUCTIONS

What amounts should be shown under
(b) Total deductions?
(c) Net wage?

NAME: Kate Brown				WORKS NO: 749		
	MORNING		AFTERNOON		HOURS	
	IN	OUT	IN	OUT	NORMAL	O/T
MON	08.00	12.01	13.00	18.00	8	1
TUES	07.59	11.59	13.01	17.00	8	
WED	08.01	12.00	12.59	19.01	8	2
THURS	08.02	12.02	13.01	17.01	8	
FRI	08.00	12.01	12.58	17.00	8	
SAT	09.01	12.00				3
SUN						
				TOTAL HOURS	40	6

SMALLWORKS ENG. LTD.   Kate Brown

	BASIC WAGE	O/T AT 1½	O/T AT 2	GROSS WAGE
PAYMENTS	76.00	17.10	−	93.10
	NAT. INS.	TAX	PENSION FUND	TOTAL DEDUCTIONS
DEDUCTIONS	14.48	13.25	1.90	
				NET WAGE

6. Design payslips for the following employees using the information given.
   (a) Rajesh Patel; Trainee; Normal hours 40; Overtime: 4 at 1½; Nat. Ins. £14·48; Tax £4·64; Pension Fund £1·40.
   (b) Carol Groves; Clerical; Normal hours 40; Overtime: 6 at 1½, 4 at 2; Nat. Ins. £14·48; Tax £17·80; Pension Fund £1·90.

7. George Wilson is an assembly worker. He is paid a basic rate of £2 per hour plus an **extra** 5p for every part that he assembles.
   Copy out and complete the table.
   The bonus pay for Tuesday was found by multiplying the parts assembled (250) by 5p.

Day	Hours Worked	Parts Assembled	Basic Pay	Bonus Pay	Total for Day
Mon	7	210			
Tues	8	250	16·00	12·50	28·50
Wed	8	240	16·00		
Thurs	8	230			
Fri	7	190			
Weekly Totals	38	1120			

23 Wages and Salaries

8. What would be George Wilsons total wage for a week in which he worked 40 hours and assembled the following parts:

Monday	270	Thursday	200
Tuesday	280	Friday	220
Wednesday	260		

9. Jim Townsend sells double glazed windows and doors. He receives a basic salary of £4200 per year plus 2% commission on goods sold. His salary plus any bonus is paid to him each calendar month.

The graph below shows the amounts of sales he makes each month.

(a) What is his **basic** salary each month?
(b) How much commision did he earn in June?
(c) What was his **total** salary for August?
(d) What was his **total** salary for March?

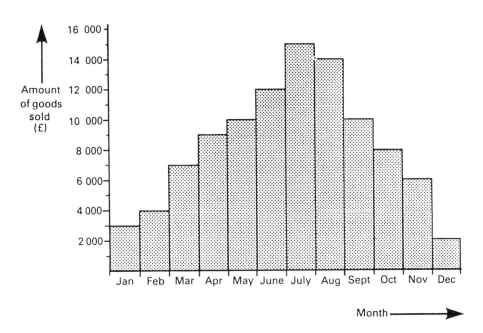

10. Bill Evans is an engineer. The company he works for offers salaries based on the length of time he works for the company up to a maximum of 15 years. Each year during this time his company gives him an increase or 'increment' to his basic salary.
    Below is his salary scale:

Starting at £4250 per year	
For the first five years	£250 increment at the end of each year.
For the next five years	£375 increment at the end of each year.
and the last five years	£425 increment at the end of each year.

    Make up an 'Incremental Table' of the type shown here and use it to find Bill Evans salary after:

    (a) six years,
    (b) eleven years,
    (c) fifteen years.

End Year	Salary
1	£4500
2	£4750
3	

For discussion

Discuss the advantages and disadvantages of being paid in the different ways mentioned in the section.

Project work

1. Find out about National Insurance.

2. Find out about income tax, i.e. what are the single persons and married persons allowances? What is the current rate of income tax?

24 Bank Accounts

More than half the working population now have their wage or salary paid directly into a **current account** at a bank. Current account holders are issued with cheque books and this saves having to carry around large amounts of cash for everyday transactions.

Account holders can arrange with their banks to have regular payments made directly from their accounts. These payments are usually referred to as standing orders (S/O's) or direct debits (D/D's).

An account is said to be 'overdrawn' (O/D) when the holder spends more money than he/she actually has in the account.

Banks send out **statements** to their customers at regular intervals.

In the **incomplete** statement below:

Payments — are amounts paid out by Mr Henry i.e. cheques, standing orders, direct debits.
Receipts — are amounts paid into the account by Mr Henry i.e. wages each month.
Balance — is the amount he has left in his account at any time.

Albion Bank — Northtown Branch

O/D · Overdrawn
S/O · Direct Debit
D/D · Standing Order

Account Number 694572

HENRY W
112 Windsor Crescent
NORTHTOWN

All entries to 29 Aug 82

Date	Particulars	Payments	Receipts	Balance
1982	Opening Balance			125 56 *
19 JLY	632913	65 00		60 56 *
26 JLY	632914	43 19		17 37 *
30 JLY	SALARY		512 75	530 12 *
1 AUG	BUILDING SOCIETY S/O	147 81		
4 AUG	632915	70 00		
5 AUG	LIFE-ASSURANCE D/D	22 20		
11 AUG	632916	45 45		
12 AUG	632917	75 00		
18 AUG	632918	39 95		
21 AUG	632919	60 00		
27 AUG	632920	11 56		

Example

How much should have been shown in the balance column on 1st August?

On first of August S/O for £147·81
So, Balance = Old balance − £147·81
 = £530·12 − £147·81
 = £382·31

Exercise 24.1

1. How much should have been shown in the balance column on
   (a) 4th August   (b) 5th August
   (c) 18th August   (d) 27th August

2. On the 29th August Mr Henry writes out a cheque for £62·50. By how much has he **overdrawn**?

3. On 31st August Mr Henry's salary is paid into his account. He receives £595·50. What amount would be shown in the balance column?

4. Make up a statement for Mr Henry from 31st August until 15th September if the following payments are made; 1st September Building Society S/O £147·81, 4th September £55·00, 5th September Life Assurance D/D £22·20, 10th September £80, 15th September £68·95.

5. On the 1st December 1982 Sue Carson opened a current account with £315·60. Make up a statement for Sue if she makes the following payments; 5th December S/O £60, 8th December £35, 10th December £29·75, 12th December £9·07, 15th December S/O £34·85, 17th December £80 and 21st December £12·35.

For discussion

What are the advantages and disadvantages of having a current account and a cheque book?

Project

Find out about other services offered to holders of current accounts.

25 Savings

People who wish to save their money place it in a deposit or savings account at a bank or building society. The bank or building society gives a payment known as **interest** to those who do this.

Insterest is paid yearly or half-yearly at a certain **percent per annum.**

Some peole **invest** a **lump sum** and have the interest paid out to them twice a year.

Examples The interest paid out on £500 invested at 10·5% for 1 year would be (using a calculator):

Press

[5] [0] [0] [×] [1] [0] [·] [5] [%] [=]   Display reads 52.5

=£52·50

and, the interest on £2000 at 8·7% for 5 years would be:

Press

[2] [0] [0] [0] [×] [8] [·] [7] [%] [×] [5] [=]

Display reads 870.

=£870

This type of interest is called **simple interest**.

Simple interest = sum invested × rate % × time

Note: The sum invested is sometimes known as the **principal**.

Exercise 25.1

Use a calculator to find the simple interest paid out in each case below. (Work to the nearest penny where necessary).

	SUM INVESTED (£)	RATE %	TIME IN YEARS
1.	1800·00	7·9	3
2.	12500·00	11·1	5
3.	8450·00	9·3	4
4.	6940·00	7·6	2
5.	7880·00	7·25	3
6.	753·00	8·9	5
7.	531·50	13·2	7
8.	8245·40	10·8	3
9.	425·00	15	$2\frac{1}{2}$
10.	3251·60	11·25	$3\frac{1}{2}$

25 Savings

Some people who invest their money do not have the interest paid out to them each year but have it added to the amount invested instead. This allows the lump sum to grow and this means that each year interest is paid on a **larger amount**. This is known as **compound interest**.

Example How much does £1500 grow to if invested for 3 years at 7·5% compound interest?

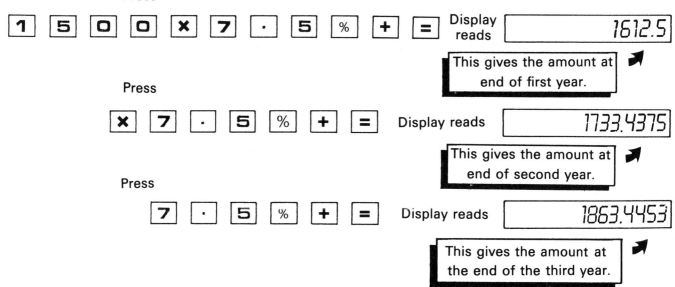

Rounded off to the nearest penny this amounts to £1863·45

Interest earned = £1863·45 − £1500 = £363·45

Exercise 25.2

Copy out the chart and use a calculator to fill in the amount and the compound interest earned (work to the nearest penny where necessary).

	SUM INVESTED (£)	RATE %	TIME IN YEARS	AMOUNT	INTEREST EARNED
1.	1500	7·5	2		
2.	2400	11·2	3		
3.	25000	9·6	3		
4.	540	10·7	4		
5.	120	11·9	5		
6.	50000	8·7	3		
7.	1255	6·9	4		
8.	1000000	15·7	3		

9. A man invested £6500 at 7·9% compound interest. After 4 years he withdrew £4000 and left the remainder for two more years. What amount did he have at the end of the two years?

10. How much **more** interest could be earned by investing £18000 at 6·8% for three years at compound interest than the same sum invested at the same rate and time at simple interest?

For discussion

What are the advantages of having a savings account with a bank, building society or post office?

Projects

1. Find out about the various saving schemes available.
2. If you were given £5000 to invest where would you place it in order to obtain the maximum amount of interest?

26 Hire Purchase

More costly items can be bought using hire purchase (H.P.). Hire Purchase involves putting down a deposit and paying the remainder of the price (called the balance) by weekly or monthly instalments.

Buying goods on hire purchase is always more expensive than paying the cost price.

Example

This TV set can be bought for £200 cash or by paying a £45 deposit and twelve monthly payments of £14·90. Find:
(a) The total H.P. price.
(b) The hire purchase charge.

(a) TOTAL H.P. PRICE
    = DEPOSIT + TOTAL PAYMENTS
    = £45 + (12 × £14·90)
    = £45 + £178·80
    = £223·80

(b) THE HIRE PURHCASE CHARGE
    = £223·80 − £200
    = £23·80*

(* THIS MEANS £23·80 EXTRA IS PAID)

Exercise 26.1

1. Find the hire purchase charges on:
   (a) A £325 washing machine which can be bought with a deposit of £65 and twelve monthly payments of £24·35.
   (b) A £285 electric cooker which can be bought with a deposit of £75 and 24 monthly payments of £10·95.

2. This stereo can be bought for £255 cash, or with a deposit of £65 and 12 monthly payments of £19·85. Find:

   (a) The total H.P. price.
   (b) How much would be saved by paying the cash price?

26 Hire Purchase

Borrowing Money

What is really happening when we buy something on H.P. is that we are borrowing money from a finance or loan company who charge interest on the amount we borrow.

Usually the finance company requires a minimum deposit and a maximum time we are allowed to pay.

Exercise 26.2

The chart below gives a finance company's requirements.

ITEM	MINIMUM DEPOSIT	MAXIMUM NO. OF MONTHS TO PAY
TVs, radio, Hi-fi, photographic equipment.	20%	30
Home Improvements, furniture, central heating.	10%	36
Washing machines, fridges, freezers, cookers etc.	20%	24
Cars, new and secondhand	20%	36
Motorcycles, boats, caravans, musical instruments.	10%	36

1. What would be the minimum deposit on the following?
   (a) £350 of photographic equipment.
   (b) A £650 3-piece suite.
   (c) A £195 deep freezer.
   (d) A £3250 car.
   (e) A £5950 caravan.

H.P. Charges

The table below shows how much has to be paid back when money is borrowed.

AMOUNT OF LOAN £	12 MONTHS monthly payment £	24 MONTHS monthly payment £	30 MONTHS monthly payment £	36 MONTHS monthly payment £
50	4·62	2·54	2·12	1·84
100	9·24	5·08	4·24	3·68
200	18·48	10·16	8·48	7·36
300	27·72	15·24	12·72	11·04
400	36·96	20·32	16·96	14·72
500	46·20	25·40	21·20	18·40
1000	92·40	50·80	42·40	36·80
2000	184·80	101·60	84·80	73·60
3000	277·20	152·40	126·60	110·40

Examples

1. If I borrow £500 over 24 months
   (a) How much do I pay back?
   (b) How much interest (hire purchase charges) do I pay?

   (a) From the tables the monthly payment is £25·40
   So, amount paid back = £25·40 × 24 = £609·60

   (b) Interest paid = £609·60 − £500 = £109·60

26 Hire Purchase

2. If you buy stereo equipment costing £500
   (a) What is the minimum deposit?
   (b) How much is the balance?
   (c) How much would you repay per month over 2 years?

   (a) Min. deposit = 20% or £500 = £100 (See the table on page 66)
   (b) Balance = £500 − £100 = £400
   (c) From the tables £400 is £20·32 per month for 24 months.

Exercise 26.3

1. Use the tables to find the monthly payments on:
   (a) £600 borrowed over 12 months
   (b) £1500 borrowed over 24 months
   (c) £2600 borrowed over 30 months
   (d) £3300 borrowed over 36 months

2. If you borrow £2500 over 36 months,
   (a) what is the monthly repayment?
   (b) how much interest do you pay?

3. If you want to spend £3000 on home improvements, put down the minimum deposit and pay over 3 years.
   (a) How much deposit is required?
   (b) How much do you borrow?
   (c) How much do you repay each month?

4. This caravan costs £3500. A man wants to put down the minimum deposit but is undecided whether to repay over two years or three years.
   (a) What is the minimum deposit?
   (b) How much does he borrow?
   (c) How much per month over two years?
   (d) How much per month over three years?

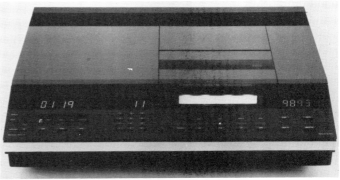

5. This video costs £540. A man pays £140 deposit and repays over two years. How much would he have saved by repaying over one year?

For discussion

What are the advantages and disadvantages of buying goods on hire purchase?

27 Value Added Tax and Service Charges

Value added tax (VAT) is a form of taxation levied by the government on most items. At the time of writing VAT is at 15%. Most items advertised for sale have VAT included in their price but certain others do not and VAT must be added to the advertised price.

Example The amount to pay for a battery marked £12·40 plus VAT would be (using a calculator):

Press

| 1 | 2 | · | 4 | 0 | × | 1 | 5 | % | + | = |

Display reads 14.26

So, the amount to pay would be £14·26.

Exercise 27.1

1. Find the amounts to pay for batteries marked at:
   (a) £17·80 plus VAT
   (b) £22·40 plus VAT
   (c) £14·50 plus VAT

2. This bill is incomplete. What amount should be shown next to:
   (a) VAT at 15%?
   (b) Total?

3. Make up a bill similar to the one shown for two tyres at £14·65 each, two valves at 90p each and two wheel balances at £1·75 each. Show VAT and TOTAL amounts on your bill.

HARRY'S tyres & batteries

Northway Road
Southtown
Tel. 65220

VAT. NO 220 334 297
CASH SALE
Date 2/2/84

1	Tyre 155 × 13		16	95
1	Valve		0	90
1	Wheel balance		1	75
			19	60
	+ VAT at 15%			
	TOTAL =	£		

27 Value Added Tax and Service Charges

Some articles have VAT included in their price

Example

This calculator and pen watch set is marked at £11·50. How much VAT is included in this price?

Marked amount = Original price + VAT

If original price = 100% and VAT = 15%, the marked amount = 115%

So $\quad\quad\quad$ 115% = £11·50

∴ 1% $\quad\quad = \dfrac{£11·50}{115}$

and VAT (15%) $= \dfrac{£11·50}{115} \times 15$

Press

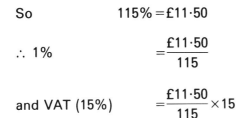

Display reads $\boxed{1.5}$

So VAT = £1·50

4. How much VAT is included in the following marked prices:
   (a) A pen marked at £3·45?
   (b) A digital watch at £10·35?
   (c) A camera at £34·50?
   (d) A pair of jeans at £11·95 (to the nearest penny)?
   (e) A skirt at £6.50 (to the nearest penny)?

5. Find the amounts paid (to the nearest penny) for items marked at:
   (a) £35·80 plus VAT
   (b) £17·95 plus VAT
   (c) £14·75 plus VAT
   (d) £125·45 plus VAT.

6. How much VAT is included in the following? (Work to the nearest penny where necessary).
   (a) A colour TV at £253
   (b) A spin dryer at £63·25
   (c) An electric cooker at £218·50
   (d) A toaster at £12·95.

7. (a) Mail order firm A offer this briefcase at £39·50 (VAT inclusive), whereas mail order firm B offer it at £35 plus VAT. Which firm is the more expensive and by how much?

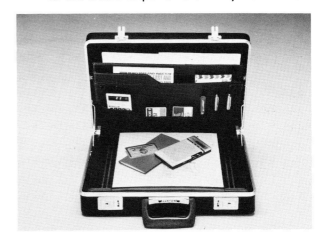

   (b) Which is the more expensive and by how much—colour television X marked at £299 (VAT inclusive) or colour television Y marked at £265 plus VAT?

27 Value Added Tax and Service Charges

Service Charges

Some hotels and restaurants add a special charge (usually 10%) for services offered.
This shows the cost of a meal for two at a restaurant. Note how the bill is totalled up and how the service charge is entered and added on.

8. Make up a bill similar to the one shown for a couple who have the following:
   Starters: 1 × 75p, 1 × 70p;
   Main course: 1 × £2·95, 1 × £3·25;
   Sweets: 1 × 65p, 1 × 85p;
   Drinks: 1 × 45p, 1 × 35p.
   Show (a) the service charge and (b) the total.

TABLE NO. 6		£
starters	1 × 45, 1 × 65	1.10
main course	1 × 1.95, 1 × 2.25	4.20
sweets	1 × 55, 1 × 65	1.20
drinks	1 × 35, 1 × 25	0.60
extras	/	
		7.10
& service (10%)		0.71
TOTAL		7.81

The County Hotel

		£	p
2	Nights B & B at £8.50	17	00
2	Dinners at £5.50	11	00
1	Lunch at £3.50	3	50
	Drinks etc.	2	50
		34	00
	+ VAT at 15%	5	10
		39	10
	Service Charge 10%	3	91
	TOTAL = £	43	01

This shows the cost of a two day stay with meals and drinks etc. at the County Hotel. Note how the bill is totalled, the VAT amount entered and finally how the service charge is entered and added on.

9. Find the total amount paid by a business man who stayed three nights and also had three dinners at £5·50 each, three lunches at £3·50 each and drinks and other items amounting to £4·50.

10. Find the total amount paid by a person who stayed at the County for 7 nights, having 6 dinners at £5·50 each, 6 lunches at £3·50 each and drinks and other items amounting to £16·50.

28 Balance Sheets

Business organisations, clubs etc. draw up **balance sheets** at regular intervals or after certain business transactions.

The purpose of a balance sheet is to show whether a **profit** or a **loss** has been made.

Shown below is the balance sheet for a youth club disco.

RECEIPTS		EXPENSES	
	£		£
Sale of tickets	140·00	Hire of hall	25·00
Sale of refreshments	60·00	DJ & equipment	40·00
		Printing tickets	12·50
		Advertising	5·75
		Cost of refreshments	42·50
Total receipts	200·00	Total expenses	125·75
Any loss	—	Any profit	74·25
Total	200·00	Total	200·00

Any profit must always be entered on the **expenses side**.

Any loss must always be entered on the **receipts side**.

$$\boxed{\text{Loss}} + \boxed{\text{Receipts}} = \boxed{\text{Expenses}}$$

Remember: the totals at the bottom of a balance sheet must always be **equal** or **balance**.

Below is an incomplete balance sheet showing the yearly accounts of a local tennis club.

RECEIPTS		EXPENSES	
	£		£
Members' subscriptions	2450	Hire of clubhouse	1560
Bar revenue	18625	Bar staff wages	4255
Sale of sports equipment	900	Buying sports equipment	850
Annual Dance	450	Refreshments for bar	12475
		Gas and Electricity	685
Total receipts	22425	Total expenses	
Any loss		Any profit	
Total		Total	

28 Balance Sheets

Exercise 28.1

1. (a) What amount should be shown for total expenses?
   (b) What amount of profit or loss was made?
   (c) What amount should be shown next to the 'total' at the bottom of both columns?
   (d) How much profit did the bar make?

2. This incomplete balance sheet shows the yearly accounts of a small firm.

RECEIPTS		EXPENSES	
	£		£
Sales of products	86850	Staff wages	75800
Sales of secondhand machinery	2900	New machinery	8500
		Advertising	1250
		Maintenance	980
		Administration	472
		Interest on loans	2548
		Gas & electricity	1820
Total receipts	89750	Total expenses	
Any loss		Any profit	
Total		Total	

(a) What amount should be shown for total expenses?
(b) What amount of profit or loss was made?
(c) What amount should be shown next to the 'total' at the bottom of both columns?
(d) The firm employs 20 staff in all. What is the 'yearly' average wage per employee?

3. Draw up a balance sheet for a local football team from the following information:

   Receipts—Members' subscriptions £110; sale of refreshments £2860; proceeds of dance £982; raffle ticket sales £400.

   Expenses—Purchasing new strips £240; cost of refreshments £1840; hire of hall £75; raffle ticket prizes £95; ground rental £1000; referees expenses £600; laundry £300; gas & electricity £180.

   What amount of profit or loss was made?

4. Draw up a balance sheet for a small engineering firm from the following information:

   Receipts—Sales of products £225,000

   Expenses—Wages £167125; new machinery £10500; Advertising £2550; maintenance £7640; administration £1125; rents £8750; gas & electricity £2560.

   What amount of profit or loss did the firm make?

29 The Holidays

Most people try to take a holiday each year either at home or abroad.
Millions travel each year to sunnier places.
The chart below shows the cost of a summer holiday in either of
two hotels on Spains Costa Brava.

HOTEL	DON JUAN								OASIS PARK							
	Available on All Flights								Available on All Flights							
Number of Nights	7		14		10		11		7		14		10		11	
Adult/Child	Ad	Ch	Ad	Ch	Ad	Ch	Ad	Ch	Ad	Ch	Ad	Ch	Ad	Ch	Ad	Ch
25 Mar. - 3 Apr.	128	100	153	100	147	107	143	102	143	122	177	143	166	139	163	134
4 Apr. - 29 Apr.	119	Free	146	Free	138	Free	135	Free	127	64	160	85	149	78	146	74
30 Apr. - 13 May	146	Free	179	Free	168	Free	165	Free	147	74	181	96	170	89	167	84
14 May - 20 May	145	57	176	57	165	60	162	58	150	77	183	97	171	90	168	85
21 May - 10 Jun.	154	89	190	88	176	94	173	86	158	115	197	141	181	131	188	127
11 Jun. - 17 Jun.	155	114	191	134	177	128	174	124	159	122	198	149	182	139	181	136
18 Jun. - 1 Jul.	158	122	203	148	183	139	182	135	163	131	213	166	190	151	190	149
2 Jul. - 15 Jul.	169	130	222	159	198	148	198	145	175	140	232	180	206	163	207	161
16 Jul. - 14 Aug.	188	158	240	187	218	177	218	174	193	168	252	209	225	192	226	191
15 Aug. - 28 Aug.	178	144	221	168	203	160	202	156	187	156	237	191	216	178	216	176
29 Aug. - 18 Sep.	165	129	204	151	189	144	187	140	170	138	215	168	196	157	195	154
19 Sep. - 25 Sep.	164	126	197	150	186	143	183	139	169	128	205	152	192	145	190	141
26 Sep. - 23 Oct.	137	95	168	117	157	109	154	112	139	93	171	113	159	107	157	109
SUPPLEMENTS per person per night	Single room 80p.								Single room 80p.							

Departures on or between

To these prices will be added: 1) Airport Charges £9.90. 2) Insurance premiums (£6.50 up to 8 days; £7.50 up to 12 days; £8.50 up to 17 days) 3)

Example A couple with one child book a 10 day holiday at the DON JUAN from the 16th July. What will be the total cost if they require a single room for the child?

				£
Charge for two adults	=	£218 × 2	=	436·00
Charge for one child	=	£177 × 1	=	177·00
Single room supplement	=	80p × 10	=	8·00
Airport charges	=	£9·90 × 3	=	29·70
Insurance premiums	=	£7·50 × 3	=	22·50
				673·20

1. (a) Between what dates does the Don Juan offer free holidays for children?
   (b) What are the most expensive dates to take a holiday?

2. What would be the total cost of the following holidays?
   (a) A couple booking a 14 day holiday at the Oasis Park from the 2nd of July.
   (b) A couple with one child booking a 7 day holiday at the Don Juan from the 19th of September. They require a single room for the child.

3. Two friends require a 7 day holiday at the Don Juan from the 9th of May.
   (a) Find the cost if they require single rooms.
   (b) How much will they save if they decide to share a room?

4. How much would a couple save by commencing an 11 day holiday at the Don Juan on 30th April instead of the same holiday commencing 15th of August?

5. How much would a couple with two children save on a 14 day holiday at the Oasis Park by commencing their holiday on 26th September instead of 16th July? (Assume that the two children would share a room.)

29 The Holidays

6. Below is information about flying times from British airports to the holiday island of Majorca.

Local time is 1 hour in front of British time i.e. 1200 in Britain is 1300 in Majorca.
Give the local times of arrival for the following flights to Palma.
(a) 14.45 from Gatwick.
(b) 16.35 from Manchester.
(c) 08.45 from Newcastle.
(d) 22.40 from Luton.

7. A flight left East Midlands at 10.30. On arrival at Palma Airport a party waited 40 minutes for a coach to take them to Magaluff. At what local time did they arrive in Magaluff?

8. A flight left Luton at 21.15 one Friday evening and was 25 minutes late on arrival in Palma due to turbulent weather. At Palma all the passengers had to wait one hour for coaches to take them to their hotels. At what local times would the passengers arrive at:
(a) Arenal? (b) Paguera? (c) Cala Bona?

9. This bar chart compares the temperatures in Majorca and London between October and April.
(a) What month has the greatest temperature difference?
(b) What is the average temperature in Majorca?
(c) What is the average temperature in London?

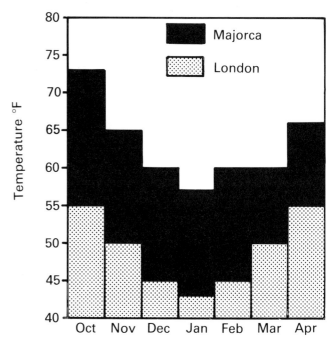

10. The chart below gives the average hours of sunshine per day in Majorca and London between October and April.

MONTH	OCT	NOV	DEC	JAN	FEB	MAR	APR
Majorca	6·2	5·1	4·4	5·0	6·0	6·5	7·4
London	3·5	2·0	1·6	1·8	2·0	4·0	5·4

(a) Find the average for Majorca.
(b) Find the average for London.
(c) Draw a bar chart (similar to the one on temperature) using the above information.
(d) What months have **above** average sunshine for London?
(e) What months have **below** average sunshine for Majorca?

For discussion What are the advantages and disadvantages of taking holidays:
(a) in this country? (b) abroad?

Project Carry out a survey among classmates and friends to find out what they did during the holidays. Find out how many:
(a) stayed at home   (b) had a camping holiday
(c) visited friends or relatives   (d) had a holiday abroad.
(e) had a holiday in a guest house or hotel in this country
Use the information you obtain to draw a bar chart.

30 Holiday Exchange Rates

The chart below shows how much in foreign currencies could be obtained for £1 sterling.

AUSTRIA	25·10	Schillings
BELGIUM	72·5	Francs
FRANCE	10·78	Francs
GERMANY	3·57	Marks
GREECE	123·00	Drachmas
ITALY	2120·00	Lira
PORTUGAL	139·00	Escudos
SPAIN	199·00	Pesetas
USA	1·49	Dollars

Exercise 30.1

Use a calculator for the following questions:

1. Change the following amounts of sterling
   (a) £50 to Marks
   (b) £120 to Dollars
   (c) £25 to Lira
   (d) £70 to French francs

2. Change the following into sterling, working to the nearest penny.
   (a) 1800 Schillings
   (b) 900 Dollars
   (c) 500000 Pesetas
   (d) 1000000 Lira

3. This price list appears for the benefit of English tourists in Spain. The prices are in pesetas. Find the equivalent sterling prices (working to the nearest penny) for:
   (a) steak pie
   (b) mixed grill
   (c) 2 pork chops
   (d) liver, bacon and onions.

4. Before going on holiday in France a family change £100 in francs. While in France they spend 936 francs.
   (a) How much do they have left?
   (b) How much will they receive when they change it back to sterling? (Work to the nearest penny)

5. This clock is marked for sale at 172 marks.
   (a) How much is this in sterling?
   (b) How much is this in drachmas?
   (c) How much is this in Belgian francs?

```
PEDRO'S      Menu

2 PORK CHOPS + PEAS *              3 25
STEAK, PEAS & SALAD *              4 00
HAM & TOMATOES *                   1 75
MIXED GRILL *                      3 50
OMELETTE & PEAS *      FROM        1 70
2 SAUSAGES *                       2 25
2 EGGS *                           1 75
2 EGGS & BACON                       85
EGGS ON TOAST                      1 10
BEANS ON TOAST                     1 25
BOILED EGG, BREAD & BUTTER           50
TOAST & MARMALADE     PER SLICE      40
HAMBURGER & PEAS *                 1 75
LIVER, BACON & ONIONS *            2 75
STEAK PIE & ONIONS *               2 50

* ALL SERVED WITH EITHER CHIPPED,
  BOILED OR MASHED POTATOES *
```

6. Find how many Spanish pesetas could be exchanged for the following currencies. (Work to the nearest whole peseta in each case.)
   (a) One Austrian schilling.
   (b) One mark.
   (c) One dollar.
   (d) One French franc.

30 Holiday Exchange Rates

7. This car can be hired for 850 escudos per day. If the car is hired for 7 days or more a reduction of 10% is given. Find, to the nearest penny, the equivalent sterling cost of hiring the car for:
   (a) 3 days
   (b) 8 days
   (c) 14 days.

8. This caravan can be hired for 35 marks per day in Germany *or* for 225 schillings per day in Austria.
   If the caravan is hired for 10 or more days in Germany a discount of 10% is given. Austria offers no discount.
   In which country would it be cheapest and by how much to:
   (a) Hire a caravan for 7 days?
   (b) Hire a caravan for 10 days?
   Give your answers in sterling.

For discussion

The exchange rates at the beginning of this section were correct at the time of writing but have probably changed by now i.e. the pound will be higher against certain currencies and lower against others. Give reasons for these changes.

Projects

1. Obtain a national newspaper with up to date exchange rates and find out whether the following currencies are higher or lower than they were.
   (a) The French franc.
   (b) The German mark.
   (c) The peseta.
   (d) The drachma.

2. Use the new exchange rates to re-do questions 4, 5, 7 and 8.

HOUSEHOLD.........

31 Buying a House

Hausman. Estate Agent
A selection of houses from this weeks NEW listings!

£ 21,500 LEASEHOLD
£100,000 FREEHOLD
£ 29,500 FREEHOLD
£ 63,000 FREEHOLD
£ 45,950 FREEHOLD

Buying their own home is the biggest financial commitment most people ever undertake. Houses are very expensive because of the high costs of land, building materials and labour.

More than half of the houses in this country are occupied by people who are buying them. Because of the high costs involved most people obtain a loan from a Building Society, bank or local council to buy their own homes. A loan given for this purpose is better known as a **MORTGAGE**.

In special cases the entire value of a property is offered for loan— this is known as a 100% mortgage.

More usually mortgages of up to 90% are offered and this means that homebuyers have to save for a deposit themselves.

Example

Jim and Sheila Brown would like to buy this house with the help of a 90% mortgage. This means they would have to find 10% of the cost themselves.
10% of £24500 = £2450
They would have to borrow £24500 − £2450 = £22050

15 Meadoways
ENTRANCE PORCH
ENTRANCE HALL
FINE THROUGH LIVING ROOM
LUXURY FITTED KITCHEN
LANDING
THREE BEDROOMS
BATHROOM WITH SEPARATE WC
AMPLE 13 AMP POWER POINTS
FULL GAS FIRED CENTRAL HEATING
DELIGHTFUL GARDENS
BRICK BUILT GARAGE

FREEHOLD... £24,500

31 Buying a House

How Much can be Borrowed?

The size of the mortgage offered to the buyer depends on his or her salary, usually up to $2\tfrac{1}{2}$ times gross annual salary. In the case of a couple (joint borrowers) up to $2\tfrac{1}{2}$ times the larger income plus once the smaller income.

Example A couple earn £6800 and £4000 per year respectively. What is the maximum mortgage they can obtain?

Maximum loan = (Larger amount $\times 2\tfrac{1}{2}$) + (smaller amount $\times 1$)
= (£6800 $\times 2\tfrac{1}{2}$) + (£4000 $\times 1$)
= £17000 + £4000
= £21000

70% Mortgage

£25,500... FREEHOLD
GOOD SIZE GARDENS

80% Mortgage

£42,500... FREEHOLD
PLEASANT SECLUDED GARDENS

Exercise 31.1

1. What deposit is required for these two houses?

2. What would be the maximum mortgages offered to:
   (a) A salesman earning £9700 per year.
   (b) A couple earning £7900 and £6500 respectively.

3. Miss Watson would like to buy this house with an 80% mortgage.
   (a) What deposit would she require?
   (b) What amount would she wish to borrow?
   (c) Could she afford it?
   Miss Watson earns £9400 per year.

5 The Green

LIVING ROOM
DINING ROOM
FITTED KITCHEN
UTILITY AREA
BATHROOM
LANDING
TWO BEDROOMS
RE-WIRED WITH AMPLE 13 AMP
   POWER POINTS THROUGHOUT
FULL GAS FIRED CENTRAL
   HEATING
MAGNIFICENT 240' GARDEN
GARAGE SPACE

FREEHOLD... £25,950

77 Suffolk Ave

ENTRANCE PORCH
ENTRANCE HALL
LIVING ROOM
DINING ROOM
FITTED KITCHEN/BREAKFAST
                                   ROOM
LANDING
3 BEDROOMS
BATHROOM AND WC
RE-WIRED
GAS FIRED CENTRAL HEATING
PLEASANT GARDENS

£31,000... FREEHOLD

4. Bob and Betty Fielding would like to buy this house with a 90% mortgage.
   (a) What deposit would they require?
   (b) How much would have to be borrowed?
   (c) Could they afford it?

Bob earns £8500 per year and Betty earns £4500 per year

31 Buying a House

5. Jack and Yvonne Scott would like to buy this house with a 70% mortgage.
   (a) What deposit would they require?
   (b) How much would they have to borrow?
   (c) Could they afford it?

   Jack earns £14800 per year and Yvonne doesn't work.

35 Fenhurst Road
ENTRANCE HALL
CLOAKROOM
LIVING ROOM
DINING ROOM
LUXURY KITCHEN/BREAKFAST ROOM
LANDING
FOUR BEDROOMS
EN SUITE SHOWER ROOM
PLUS BATHROOM
GARAGE
PLEASANT GARDENS
CENTRAL HEATING

£45,500

Mortgage Repayments

Mortgage repayments have to be paid each month over 15, 20 or 25 years.

The table below shows the amounts to be paid.

SIZE OF LOAN £	OVER 15 YEARS	OVER 20 YEARS	OVER 25 YEARS
100	0·90	0·84	0·81
200	1·80	1·68	1·62
300	2·70	2·52	2·43
400	3·60	3·36	3·24
500	4·50	4·20	4·05
1000	9·00	8·40	8·10
2000	18·00	16·80	16·20
3000	27·00	25·20	24·30
4000	36·00	33·60	32·40
5000	45·00	42·00	40·50
10000	90·00	84·00	81·00

Example A couple borrow £15 900 over 25 years. What will be their monthly repayments?

From the table:

		£	
10000	requires	81·00	repayment per month
5000	requires	40·50	repayment per month
500	requires	4·05	repayment per month
400	requires	3·24	repayment per month

So £15900 requires £128·79 repayment per month

31 Buying a House

Exercise 31.2

1. What would be the monthly repayments on:
   (a) £17200 borrowed over 15 years?
   (b) £19800 borrowed over 20 years?

2. Use a calculator to find the total amounts to repay on:
   (a) £22000 over 25 years.
   (b) £19200 over 15 years.

3. A couple borrow £10000 over 25 years. Use a calculator to find:
   (a) The total amount they would have to repay.
   (b) How much less would they repay if they borrow the same amount over 20 years.

4. A couple intend to buy this house with an 80% mortgage over 25 years.

63 Offord Place
ENTRANCE PORCH
ENTRANCE HALL
CLOAKROOM
LIVING ROOM
DINING ROOM
FITTED KITCHEN/BREAKFAST ROOM
LANDING
4 BEDROOMS
BATHROOM & WC
FULL GAS FIRED CENTRAL HEATING
AMPLE 13 AMP POWER POINTS
PLEASANT SECLUDED GARDEN
CARPETS AVAILABLE
GARAGE

£ 42,500   FREEHOLD

   (a) How much do they need to borrow?
   (b) What will be their monthly repayment?

For discussion

1. What are the advantages and disadvantages of:
   (a) buying or renting a house
   (b) buying old or new property?

2. Why do property prices tend to vary from area to area?

Projects

1. Most people sell their houses through an estate agent who charges commission for this service. Obtain a copy of your local paper and looking under the property for sale section find out how much estate agents charge (usually a percentage of the price).
Use the local paper to find the cheapest and most expensive properties for sale and calculate how much the estate agent will charge in each case.

2. Find out about other expenses involved in buying or selling houses.
   (a) Stamp duty.
   (b) Solicitors fees.

32 Household Insurance

Insurance is a form of protection against loss or damage caused by fire, explosion, vandalism etc. It also covers the cost of replacing household contents. The insurance company charges the person taking out the insurance policy an amount known as the **premium** each year.

Additional insurance cover is available for:

Freezer contents— in the case of failure due to defects, failure of electricity supply etc.
Television — loss of or damage to a set.
Jewellery — cameras and photographic equipment, stereos, musical instruments.

Examples of premiums required to insure buildings and their contents are listed below.

Items	Premium
Buildings	£1·50 per £1000 value
Contents	£5·50 per £1000 value
Deep Freezer contents	£4·50 per £100 value
Televisions & Video recorders	£2·50 per colour set
Jewellery Sports equipment, stereos, musical instruments	£6·50 per £100 value
cameras etc.,	£3·00 per item

Example A family wish to insure their £29000 home. The contents are valued at £5000 and additional items to be insured are deep freezer contents £200, one colour TV set and one music centre. What will be the total premium due?

Item	Cost	£
Building (£29000) =	29 × £1·50 =	43·50
Contents (£5000) =	5 × £5·50 =	27·50
Freezer contents =	2 × £4·00 =	8·00
Colour TV =	1 × £2·50 =	2·50
Music centre =	1 × £3·00 =	3·00
	Total =	84·50

Exercise 32.1
Find the total cost of insuring each of the following buildings and their contents.

1.
ITEM	VALUE
Flat	£20000
Contents	£3000
One colour TV	
One video	

2.
ITEM	VALUE
House	£24000
Contents	£4000
Freezer contents	£100
One colour TV	
One camera	

3.
ITEM	VALUE
Bungalow	£30000
Contents	£5000
Freezer contents	£200
Jewellery	£1000
Colour TV	
One piano	

4.
ITEM	VALUE
Flat	£34000
Contents	£6000
Freezer contents	£200
Colour TV	
Video	
Jewellery	£400

5.
ITEM	VALUE
House	£38000
Contents	£6000
Two colour TVs	
Video	
Jewellery	£700
Set of golf clubs	

6.
ITEM	VALUE
Bungalow	£38000
Contents	£5000
Freezer contents	£200
Jewellery	£2000
Two cameras	
One colour TV	

7.
ITEM	VALUE
House	£62000
Contents	£10000
Jewellery	£4000
Two colour TVs	
One video	
One stereo	
One piano	
Deep freezer contents	£300

8.
ITEM	VALUE
House	£95000
Contents	£18000
Jewellery	£5000
Three colour TVs	
Two videos	
Two stereos	
Two sets of golf clubs	
Deep freezer contents	£400
Two cameras	
Three musical instruments	

33 The Rates

The rates are a form of taxation levied by the local, borough, metropolitan or county council to pay for services they provide such as education, police, fire-brigade, refuse collecting etc.

The pie chart below shows how Northtown Borough Council plan to spend the £18 million they will collect next year from the rates.

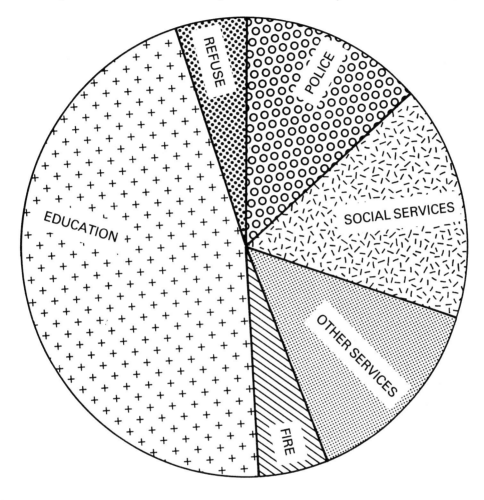

Exercise 33.1

1. What scale is the pie chart using? (i.e. 1°=£?)

2. Use a protractor to find the angles that represent:
   (a) the police
   (b) social services
   (c) education.

3. What amount is to be spent on:
   (a) education
   (b) the police
   (c) other services?

4. What percentage of the total amount is to be spent on:
   (a) the police
   (b) refuse collection?

5. Neighbouring Southtown Borough Council need to collect £54 million next year and intend to spend it as follows:

Education	£24 million
Refuse Collecting	£3 million
Police	£6 million
Fire-brigade	£3 million
Social services	£12 million
Other services	£6 million

   Construct a pie chart to show the above information.

33 The Rates

How Rates are Worked Out

Each year the council estimate how much they will need.

Occupiers of houses have to pay according to the rateable value of the property they live in. The more valuable the property the higher the rate.

Occupiers who pay rent often have their rates included in their rent.

To arrive at the rate the council estimates how much it will need in the next year and divides that amount by the rateable value of all the properties in the area.

Example Northtown Council's estimated expenditure is £18 million and the rateable value of all the properties is £15 million. Find the rate.

$$\text{Rate} = \frac{\text{Estimated Expenditure}}{\text{Rateable Valuable}}$$

$$= \frac{£18 \text{ million}}{£15 \text{ million}} = £1.20$$

OR WE SAY 120p in the £

Exercise 33.2

1. A town council has an estimated expenditure of £25 million and the rateable value of all its properties is £20 million. Find the rate.
2. Find the rate of a town which has an estimated expenditure of £13,500,000 and a total rateable value of £10,000,000.
3. Find the rate of another town which has an estimated expenditure of £32,500,000 and a rateable value of £25,000,000.
4. The rate of a town is 150p in the pound. Find its rateable value if its estimated expenditure is £42,000,000.
5. The rateable value of all the properties in a town is £12 million. Find the estimated expenditure for next year when it is planned to have a rate of 135p in the pound.

How the Occupiers Pay

Occupiers who pay rent usually have the rates included in their rent.

People who own or are buying their house with a mortgage are given the opportunity to pay their rates either yearly, half yearly or by **ten** monthly instalments.

Example Mr Smith lives in a town where the rates are 120p in the pound. The rateable value of the house he owns is £250.
    (a) How much in rates does he pay each year?
    (b) If he elects to pay by means of ten monthly instalments, how much will he pay each time?
    (a) Rates paid = rateable value × rate
                 = 250 × 120p = 30000p
                  = £300 per year

    (b) By instalments: $\frac{£300}{10} = £30$

Exercise 33.3

1. Another family in the same town as described in the example have a property with a rateable value of £285. How much will they pay each time if they elect to pay by the instalment method?
2. A town has a rate of 186p in the pound. How much will be the yearly amount paid for a house with a rateable value of £220?
3. A man pays £240 per year rates. The town he lives in has a rate of 120p in the pound. Find the rateable value of his property.
4. A property has a rateable value of £225. The owners pay £360 in rates each year. What is the rate?
5. A certain town which has a rate of 190p in the pound offers a 10% discount if the full amount is paid promptly at the start of each year. How much could be saved by paying this way on a property with a rateable value of £250?

For discussion
Discuss the ways that councils spend their rate money.
Projects
1. Find out the difference between local or district and metropolitan borough or county councils.
2. Find out the rates in your own and neighbouring areas.

34 The Electricity Bill

Electrical energy is produced at power tations, transmitted at high voltage by means of overhead cables, which are supported by steel towers or pylons. The overhead cables end at buildings known as substations where the high voltages are transformed down and fed to factories and homes by means of underground cables. As it enters factories and homes the supply passes through a meter which records the number of units of electrical energy used.

Every quarter (13 weeks) the meter is read to find how many units have been used.

Some of the meters are of a digital type and look like this:

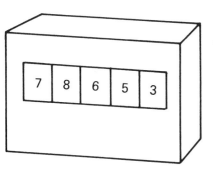

A This shows the meter reading at the **beginning** of a quarter.
B This shows the meter reading at the **end** of a quarter.

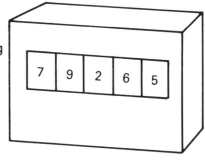

To find the number of units used subtract **A** from **B**

i.e. 79265
    −78653
    ―――
     612

If the meters shown had belonged to the Brown family they would have received this bill.

ANGLIAN ELECTRICITY at your service

Mr C Brown
21 Priory Close
Southtown

METER READING		UNITS USED	PENCE PER UNIT	AMOUNT £		STANDING CHARGE		VAT CODE	TOTALS £	
THIS TIME	LAST TIME									
79265	78653	612	4.83	29	56	6	37	0	35	93

YOUR CUSTOMER NUMBER	YOU CAN PHONE US ON	PERIOD ENDING	AMOUNT DUE NOW
21 4751 0870 60		13 Mar 84	35.93

C against a meter reading means it is your own reading

E means the meter was not read; you should check this against your meter but any difference will be put right at the next reading

VAT 0=Zero rated   X=Exempt
A=Any VAT included in this charge will have been detailed previously
Registration Number 123 4567 89

34 The Electricity Bill

The **amount** = units used × pence per unit
= 612 × 4.83p
using a calculator = 2955·96p = 2956p (nearest penny)
= £29·56

The **standing charge** is a fixed amount that must be paid every quarter whether or not any electricity is used.

The **amount due now** is found by adding the cost of the units used to the standing charge.

Exercise 34.1

1. Below are the readings for the Smith family.

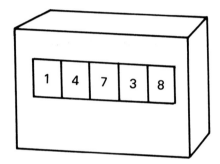

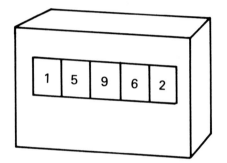

This shows the reading at the start of a quarter.

This shows the reading at the end of the quarter.

  (a) How many units have they used?
  (b) What would be the amount due if the standing charge was £6·37 and the unit rate was 4·83p?

2. Some houses have clock type meters. Some of the clocks are read clockwise and the others anti-clockwise.
   To read this kind of meter, note the number that the pointer has just passed.

Example

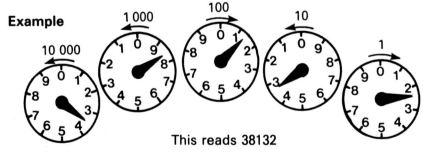

This reads 38132

Below is the Douglas family's meter at the start of a quarter.

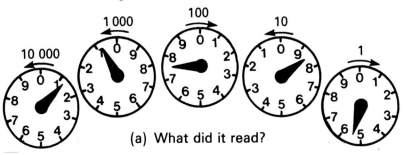

  (a) What did it read?

34 The Electricity Bill

This is the reading at the end of the quarter.

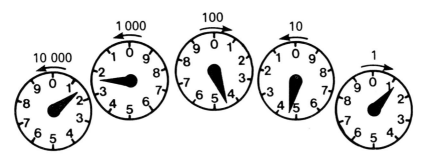

(b) What does this read?
(c) How many units have been used?
(d) Calculate the amount due if the standing charge is £6·37 and the unit rate is 4·83p.

3. Assume a standing charge of £6·37 and a unit rate of 4·83p to calculate the amounts due for families who use
   (a) 1612 units
   (b) 2422 units
   (c) 1901 units.

4. Find the amount paid per year by a family who use the following units:

1st quarter	665 units
2nd quarter	1024 units
3rd quarter	1528 units
4th quarter	779 units

   Assume a quarterly standing charge of £6·37 and a unit rate of 4·83p.

5. **Running costs:** One unit is consumed when electricity is used at the rate of one kilowatt (1000 watts) for one hour. For one unit:

This 100 W bulb will burn for 10 hours.

This refrigerator will operate all day.

This 2 kW fan heater will provide $\frac{1}{2}$ hours warmth.

Use a calculator, rounding off to the nearest penny the following costs. (Assume 1 unit costs 4·83p)
(a) Burning a 100 W bulb for 10 hours per day for 7 weeks.
(b) Operating a refrigerator for one year.
(c) Running a 2 kW heater for two hours each day for 65 days.

34 The Electricity Bill

6. This cooker uses 25 units per week in cooking a (family's) meals. Find the cost per year (52 weeks, 4·83p per unit).

7. Food freezers use $1\frac{1}{2}$ units per cubic foot capacity per week. Find the costs of using the following freezers (4·83 per unit)
  (a) 8 cubic foot for 1 week.
  (b) 15 cubic foot for 26 weeks.

8. This tank will provide hot water for a family. If well lagged the tank will use 85 units per week. Find the yearly (52 week) cost at a unit rate of 4·83p.

For discussion

Discuss the safe use of electricity in the home.

Projects

1. Find out how a fuse works.
2. Find out how an electric light bulb works.

35 The Gas Bill

The majority of homes in this country are now supplied with natural gas. Most of our natural gas is now brought ashore by means of pipes from offshore rigs. Underground pipes bring the gas to our homes. Like the electricity supply the gas flows through a meter which is read every quarter to find how many **hundreds of cubic feet** of gas have been used.

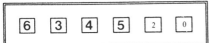

With this digital type only the four left hand numbers are read (i.e. 6345). The two smaller right hand numbers we ignore.

To find the number of cubic feet of gas used in a quarter we subtract the previous reading from the present reading.

Example   If the previous reading on the meter had been 6115 (hundreds of cubic feet).

Then consumption would be   6345
                                        −6115
                                          230   (hundreds of cubic feet)

Some Houses have Clock Type Meters

Ignore the top two clocks and read the **bottom four** from **left to right**. This one reads 3654.

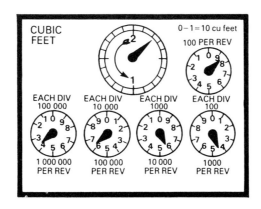

35 The Gas Bill

Exercise 35.1
1. What are the readings on these clocks?

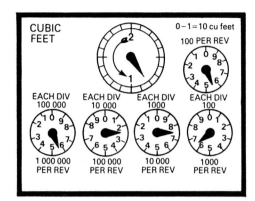
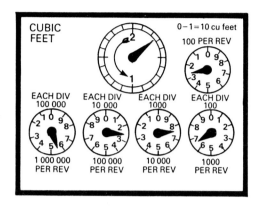

The gas bill consists of a standing charge plus unit charges for the number of **therms** used.

Therms are a measure of the quantity of heat given off when gas is burnt.

Although the meters register the number of hundreds of cubic feet of gas used the householder is charged for the number of **therms** used.

```
EAST ANGLIA GAS          VAT REGISTRATION NUMBER  123 4567 89                                              03791
TOWER POINT,             ACCOUNT REFERENCE NUMBER   DATE OF ACCOUNT          G Wilson Esq
HILLS ROAD               35 1935 0870 33 LA         02/08/84                 10 North Road
TELEPHONE 543210         PLEASE QUOTE IN ALL ENQUIRIES  (TAX POINT)          Newtown
```

	READING OR INVOICE DATE	METER READING		GAS SUPPLIED		PRICE PER THERM	AMOUNT	VAT CODE	VAT CHARGES
		PRESENT	PREVIOUS	CUBIC FEET (HUNDREDS)	THERMS				
	28 7	4148	4028	120	124.080	30.500	37.84	A	0.00
		STANDING CHARGE				9.00		A	0.00
		TOTAL VAT CODE A @ ZERO %				0.00			0.00
		TOTAL VAT							

FOR OFFICE INFORMATION				CALORIFIC VALUE	TARIFF		AMOUNT NOW DUE
N 442N	583N	163N	129N	38.6 MJ/m³ 1034 B.t.u./ft³	DOMESTIC CREDIT	£ 46.84	

If we know the number of hundreds of cubic feet of gas used, then we can calculate the therms.
The reading we obtain is multiplied by the calorific value (an energy value) and then divided by 1000 to convert it to therms.

$$\text{THERMS} = \frac{\text{NUMBER OF HUNDREDS OF CUBIC FEET} \times \text{CALORIFIC VALUE}}{1000}$$

(Calorific value = 1034)

On the bill shown, (using a calculator)
$$\text{Therms} = \frac{120 \times 1034}{1000} = 124 \cdot 080$$

35 The Gas Bill

2. Convert these readings into therms (calorific value = 1034).
   (Use a calculator)
   (a) 421 hundreds of cubic feet
   (b) 354 hundreds of cubic feet
   (c) 567 hundreds of cubic feet

3. Below are a family's meter readings at the start and end of a quarter.

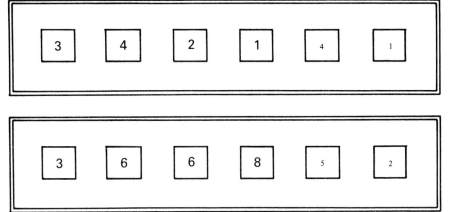

   (a) How many hundreds of cubic feet of gas have been consumed?

   (b) How many therms does this represent? (Calorific value = 1034)

   (c) Calculate the amount due if the standing charge is £9·00 and each therm costs 30·5p.

4. Another family have clock type meters.

   (a) How many hundreds of cubic feet of gas have been consumed?

   (b) How many therms does this represent? (Calorific value = 1034)

   (c) Calculate the amount due if the standing charge is £9·00 and each therm costs 30·5p.

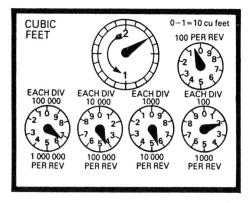

5. Find the amounts due to families who consume the following quantities of gas.
   (a) 320 hundreds of cubic feet.
   (b) 197 hundreds of cubic feet.
   Assume calorific value = 1034, standing charge = £9·00 and a charge of 30·5p per therm.

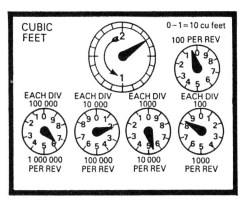

For discussion

Compare the advantages and disadvantages of gas and electricity in the home.

Projects

Find out from your classmates and friends:
(a) The number who cook with gas and electricity.
(b) What form of heating they have in their homes.

36 Telephone Charges

The telephone service is mainly provided by British Telecom.
- Customers rent the telephone and pay a rental charge plus the cost of all outgoing calls each quarter.
- The cost of calls varies according to the time of day and the distance involved.
- The chart below shows the approximate costs involved in making direct dialled (STD) calls at the time of writing.

Length of call	PEAK RATE Mon–Fri 9 a.m.–1 p.m.			STANDARD RATE Mon–Fri 8 a.m.–9 a.m. 1 p.m.–6 p.m.			CHEAP RATE All other times		
	Local	Under 56 km	Over 56 km	Local	Under 56 km	Over 56 km	Local	Under 56 km	Over 56 km
3 mins	10p	30p	75p	10p	20p	55p	5p	10p	20p
5 mins	20p	50p	1·25	15p	35p	95p	5p	15p	30p
10 mins	35p	1·00	2·50	25p	70p	1·90	10p	20p	40p

Examples The cost of making a 5 minute call to an area **more than 56 km** away at 10 a.m. on a Tuesday (peak rate) would be £1·25.

The cost of making a 10 minute local call at 2 p.m. on a Thursday (standard rate) would be 25p.

The cost of making a 3 minute call to an area **less than 56 km** away at 8 p.m. on any day (cheap rate) would be 10p.

Exercise 36.1

1. Find whether calls made at the following times would be cheap, standard or peak rates.
   (a) 4 p.m. on a Saturday.
   (b) 8.30 a.m. on a Tuesday.
   (c) 10.15 p.m. on a Wednesday.
   (d) 4.15 p.m. on a Friday.
   (e) 7.30 a.m. on a Thursday.

2. Find the costs of making the following calls:
   (a) A 10 minute local call at 10 a.m. on a Sunday.
   (b) A 3 minute call to a distance over 56 km at 10.15 a.m. on a Tuesday.
   (c) A 5 minute call to a distance under 56 km at 3.15 p.m. on a Thursday.
   (d) A 10 minute call to a distance over 56 km at 5.45 p.m. on a Saturday.

3. How much could have been saved if the following calls had been made at cheap rate times?
   (a) A 3 minute call at 12.15 p.m. on a Monday to an area **more than** 56 km away.
   (b) A 10 minute call at 4.30 p.m. on a Friday to an area **less than** 56 km away.

4. How much could have been saved if the following peak rate calls had been made at **standard rate** times?
   (a) A 3 minute call at 10 a.m. on a Tuesday to an area **more than** 56 km away.
   (b) A 10 minute call to an area **less than** 56 km away at 12 p.m. on a Friday.

36 Telephone Charges

5. This map has a scale of 1 mm = 1 km. The circle drawn around Routledge indicates local calls. Use a millimetre ruler to find which towns are
   (a) Less than 56 km from Routledge.
   (b) More than 56 km from Routledge.

6. Use the map and the chart showing telephone costs to find the cost of the following calls from Routledge.
   (a) A 3 minute call to Penton at 8 p.m. on a Friday.
   (b) A 5 minute call to Ashbridge at 10.30 a.m. on a Monday.
   (c) A 10 minute call at 3.10 p.m. on a Wednesday to Sutton.

7. Use the map and the chart to find the following costs.
   (a) A 10 minute call from Sutton to Newtown at 9.50 a.m. on a Thursday.
   (b) A 5 minute call from Penton to Ashbridge at 4.10 p.m. on a Monday.
   (c) A 3 minute call from Newton to Penton at 11.45 p.m. on a Friday.

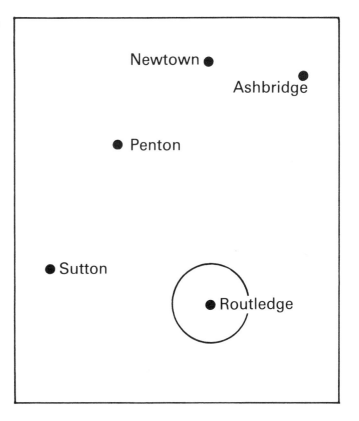

8. The chart below shows the cost of operator connected calls.

TYPE OF CALL	CHARGE RATE	FOR FIRST 3 MINUTES	EACH ADDITIONAL MINUTE
Local calls	Cheap	21p	7p
	Standard	21p	7p
	Peak	24p	8p
Up to 56 km	Cheap	28p	9p
	Standard	41p	14p
	Peak	52p	17p
Over 56 km	Cheap	41p	14p
	Standard	76p	25p
	Peak	97p	32p

Use the chart to find the cost of the following operator connected calls.
   (a) A 5 minute local call at standard rate.
   (b) A 7 minute call under 56 km at cheap rate.
   (c) A 12 minute local call at peak rate.
   (d) A 9 minute call over 56 km at peak rate.
   (e) A 20 minute call over 56 km at cheap rate.
   (f) A 15 minute call under 56 km at standard rate.

37 D.I.Y.

After buying and insuring a home, paying rates and day to day living expenses a family need to keep their home in good order.

The costs of repairs, replacing furniture and decorating can be quite high. Most families do their own decorating. Before starting to decorate a good idea is to make a plan of the areas involved. This will help to give an idea of the materials required and the cost involved.

Exercise 37.1

Shown below is a scale drawing of the downstairs of a house. The scale is 1 cm to 1 metre.

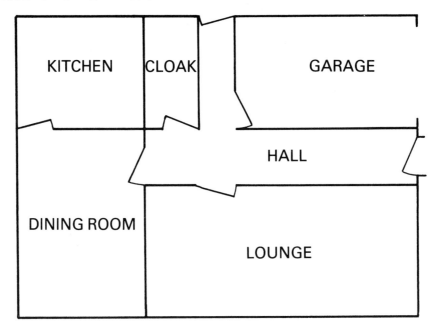

By measurement the length is 11 cm. This means the real length is 11 metres.

1. Find:
   (a) the real width of the house,
   (b) the real area of the house.

2. Find the real areas of:
   (a) the dining room,
   (b) the lounge.

The family who live in the house plan to decorate their lounge and dining room and need to buy quantities of the materials shown below.

Material	Cost
Wallpaper pattern X	£4·50 per roll
Wallpaper pattern Y	£3·75 per roll
Matt Vinyl Emulsion	£1·20 per litre
Gloss paint	£1·95 per litre
Wallpaper paste	£0·66 per packet

37 D.I.Y.

3. **The Dining Room.** The family intend to paint the ceiling with Matt Vinyl emulsion, decorate all the walls with wallpaper pattern Y and gloss paint the doors, window frames and skirting boards.
The rolls of wallpaper are 50 cm wide and after allowing for doors and windows and matching up they will get **four** drops or lengths **per roll**.
1 litre of Matt Vinyl emulsion covers 17·5 m².

   (a) What is the area of the ceiling?
   (b) How much Matt Vinyl emulsion would be used if they intend to apply **two** coats?
   (c) How many 'drops' or lengths of wallpaper will they use?
   (d) How many **rolls** of wallpaper will they use?
   (e) Find the total cost if in addition they intend to use 1 litre of gloss paint and two packets of wallpaper paste.

4. **The Lounge.** Again the ceiling is to be painted with Matt Vinyl emulsion and the walls are to be papered with **wallpaper pattern X**. Doors, windows and skirting boards are to be painted with gloss paint. **Again**, the wallpapers are 50 cm wide and they will get four drops per roll.
1 litre of Matt Vinyl emulsion covers 17·5 m².

   (a) How many litres of Matt Vinyl will be used if two coats are to be applied?
   (b) How many drops of wallpaper pattern X will they use?
   (c) How many rolls of wallpaper will they use?
   (d) Find the total cost if in addition one litre of gloss paint and three packets of wallpaper paste are to be used.

5. The family are thinking about re-carpeting the area consisting of the dining room, lounge and hall.

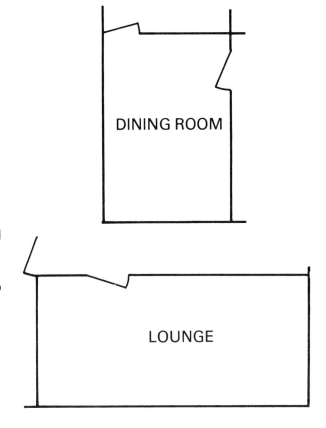

They were undecided whether to buy carpet costing £6·95 per m² plus underlay at £2·75 per m² **or** carpet tiles.
The carpet tiles are sold in boxes of 10, each tile measuring 50 cm by 50 cm. The price of the carpet tiles is £27·50 per box.
Ignore the thickness of the walls to find:
(a) The total area to be carpeted.
(b) The total cost of buying carpet and underlay.
(c) How many carpet tiles would be required?
(d) How many **boxes** of carpet tiles would they have to buy?
(e) Which method would be the cheapest and by how much?

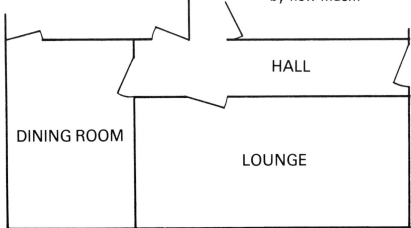

CARS . . . . →

38 Buying Cars

New Cars

New car prices consist of a list price and a special car tax. VAT at 15% is added to both list price and special car tax to arrive at the recommended retail price (the actual selling price).

MODEL	LIST PRICE	SPECIAL CAR TAX	VAT	MAX. REC. RETAIL PRICE
Metro 1·0 City	2706·00	232·72	440·81	3379·53
Metro 1·3 L	3488·00	290·67	566·80	4345·47

The above table shows how the recommended retail prices of the Metro City and the 1·3 L are arrived at. (The prices were calculated to the nearest penny.)

Recommended retail price = List price + Special car tax + VAT.

Exercise 38.1

The incomplete table below shows some prices in the Maestro range.

MODEL	LIST PRICE	SPECIAL CAR TAX	VAT	MAX. REC. RETAIL PRICE
1·3 L	3977·00	304·67	642·25	
1·3 HLE	4290·00	357·50		
1·6 L	4194·00	349·50		
1·6 HLS	4487·00	373·92		

Use a calculator to answer the questions below. (Work to the nearest penny where necessary.)

1. (a) The maximum recommended retail price for the Maestro 1·3 L.
   (b) How much VAT is charged on the Maestro 1·3 HLE?
   (c) The maximum recommended retail price for the Maestro 1·3 HLE.

2. How much VAT is charged on:
   (a) The Maestro 1·6 L?
   (b) The Maestro 1·6 HLS?

3. How much more for a Maestro 1·6 HLS than a 1·6 L?

38 Buying Cars

On the road prices

To put a new car on the road involves more than just paying the recommended retail price. Extras include: delivery charges, number plates and road tax. Additional extras that a buyer may require such as fitted radio metallic paint finish, sliding sunroof etc. are referred to as 'optional' or factory fitted extras.

Exercise 38.2

1. What would be the 'on the road price' for a Maestro 1·3 L if the following are to be added?
Delivery charge £140; Number plates £25; Road Tax £90; Metallic finish £76; Rear Window wash/wipe £55; Tinted glass £42.

2. What would be the 'on the road price' for a Maestro 1·6 HLS if the following items are to be added?
Delivery charge £140; Number plates £25; Road Tax £90; Radio/stereo cassette £139·99; Sliding sunroof £258·99; Alloy wheels £215; Metallic paint £76.

New or used cars can be bought either by paying the cash price or by hire purchase.

Exercise 38.3

1. The 'on the road price' of this new Renault 5 is £3200. The minimum deposit required is 20% and the maximum time to pay the balance is 3 years.
   (a) How much is the minimum deposit?
   (b) What would be the total price of buying this car if the minimum deposit was put down and the balance was paid off at £92·35 per month for 3 years?
   (c) How much would be saved by paying cash?

2. A man wishes to part exchange an older car as a deposit against the new Renault 5. The dealer offers him £1050.
   (a) How much extra will he have to find?
   (b) He decides to borrow the balance and pay it back over two years at £109·28 per month. What would be the total amount he will pay for the car?
   (c) How much would he save by paying cash?

3. This dealer offers 10% off his listed prices if no part exchange is offered. What would be the prices of:
   (a) The Allegro and, (b) The Rover
   to customers offering no part exchange?

HARRY'S USED CARS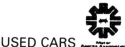

1980 AUSTIN Allegro Series 3 1.3 2-door Saloon. One owner, 15,000 miles..£2,750
1980 (V) FIESTA 1.3 Ghia. Solak gold, chocolate trim, 1 owner. 24,000 miles. Usual Ghia refinement...........................£3,395
1978 CHRYSLER Horizon GL 1.1 5-door Saloon. 33,000 miles, radio...£2,450
1977 model (R) CORTINA 1.6GL. Beige with black cloth trim, push button radio, MOT £1,795
1979 (V) RENAULT 18TS. Metallic blue with blue velour trim. Sun roof..£2,295
1977 (R) CORTINA 1.6L. Pale blue with blue trim...............£1,395
1972 (K) ROVER 3.5 Saloon Auto. Birch grey with dark tan leather interior. 48,000 miles with full service history.........£1,350
1979 (T) BEDFORD Chevanne. White........................ £1,050+VAT

ALL WITH 12 MONTHS WRITTEN GUARANTEE

---

4. A man offers his Ford Cortina as part exchange against the Fiesta.
The dealer offers £875.
   (a) How much extra will he have to find?
   (b) If he adds £600 to the deposit how much does he need to borrow?
   (c) If he borrows the remainder and pays it back at £96·58 per month for 2 years what would be the total amount paid for the car?

5. What would be the total price of the Bedford Chevanne:
   (a) If a part exchange is offered?
   (b) If no part exchange is offered?

For discussion

Buying a used car can be risky. What are the advantages and disadvantages of buying a used car privately or from a garage?

Projects

Many motorists join an organisation such as the AA (Automobile Association) or the RAC (Royal Automobile Club). Find out about the services offered by these organisations.

39 Depreciation and Running Costs

Most cars **decrease** in value as they get older.

Example What would be the value of a £4200 car in 3 years time if the rate of depreciation is 8%?

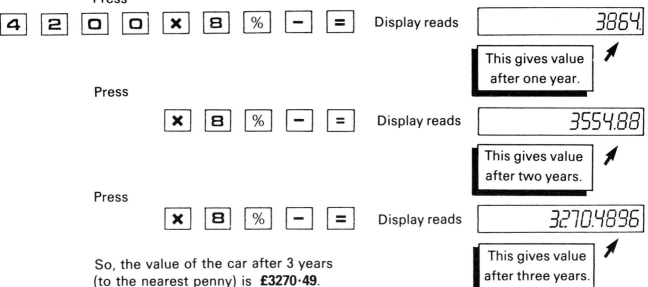

So, the value of the car after 3 years (to the nearest penny) is **£3270·49**.

Exercise 39.1
Use a calculator to the following questions.

1. A certain make of car costs £5000 today and its depreciation is reckoned at 9% per annum. How much will the car be worth in:
   (a) 3 years?   (b) 5 years?

2. A certain make of foreign cars are reckoned to depreciate at 15% per annum. Find the value of:
   (a) A £3500 car in three years.
   (b) A £6000 car in four years.

3. A car costing £5600 is reckoned to depreciate at 12% per year for three years and then 25% per year. How much will the car be worth after six years?

4. A £5000 car is reckoned to depreciate at 14% per year. How many years will it take for the value of the car to drop below £2500?

5. A certain make of car costing £9000 is reckoned to depreciate at 10%. A second make costing £8000 is reckoned to depreciate at 5%. How many years does it take for the second car to become more valuable than the first car?

Running Costs

After buying a car more expense is involved in maintenance and other costs. The largest day to day expense is buying petrol. Savings can be made by choosing the right car and by driving at economical speeds. Although garages now sell petrol in litres most people still tend to 'think' in gallons.

39 Depreciation and Running Costs

Exercise 39.2

1. This graph converts miles per gallon to litres per 100 km. (The continental method of measuring petrol consumption.)
   From the graph 40 m.p.g. is the same as 7 litres per 100 km.
   Use the graph to convert
   (a) 30 m.p.g. to litres per 100 km.
   (b) 20 litres/100 km to m.p.g.
   (c) 55 m.p.g. to litres/100 km.
   (d) 15 litres/100 km to m.p.g.

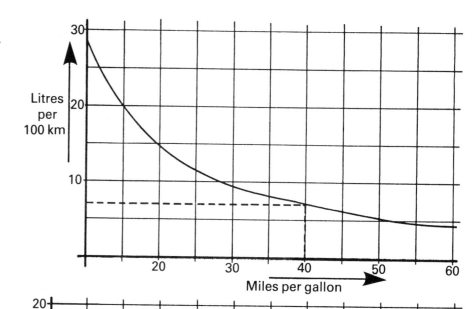

2. This graph shows the petrol consumption of a car when driven at various average speeds.
   Find how many litres per 100 km the car uses when driven at the following average speeds.
   (a) 60 km/hour.
   (b) 120 km/hour.
   (c) What is the most economical speed to drive at?
   (d) How many litres of petrol would be needed for a journey of 600 km if the average speed was 120 km/h?

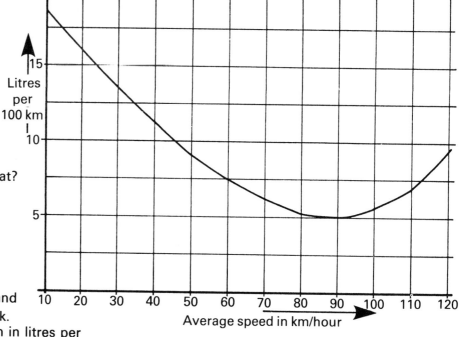

3. A family car averages 400 km and uses 32 litres of petrol per week.
   (a) Find the petrol consumption in litres per 100 km.
   If petrol costs 37·5p per litre, use a calculator to find how much is spent on petrol in:
   (b) One week.   (c) One year.

4. A company keep a record of the yearly running costs of one of their cars.
   The car travels 63000 km and has a petrol consumption of 7 litres per 100 km.

CAR EXPENSES	£
Tax & Insurance	280·50
Maintenance & repairs	265·75
Petrol	
Total	

   If the petrol costs 37·5p per litre use a calculator to find the amount that should be shown next to:
   (a) Petrol   (b) Total
   (c) Find the average weekly costs (52 weeks).

For discussion
What steps should an owner take to try and keep his/her car in good order?

Projects
1. Find out the effects on a car when it is being driven at:
   (a) Low speeds in low gear.
   (b) High speeds.

2. Make up a list of factors involved in the driving and maintenance of cars which will ensure good petrol consumption.

40 Car Insurance

It is required by law that any motor vehicle must be insured before it is used on the road.

Terms used by Insurance Companies

First party: The policy holder or a named driver i.e. a relative.
Second party: The insurance company.
Third party: Anyone else making a claim.
The Premium: The amount paid to insure a vehicle.

Types of Car Insurance

Third party only: If your car or person is damaged by some other car and the other driver is at fault claims can be made against his/her insurance policy.
Third party, fire and theft: Similar to third party only except that the driver pays extra to insure against fire and theft.
Comprehensive insurance: This covers against third party risks and damage to the drivers own vehicle.

Car insurance premiums are worked out by means of a points system.
The type and age of a car, the age of its driver and the way in which the car is to be used determine the amount to pay.
The larger the total of points, the higher the premium.
Cars are divided into **groups** for insurance purposes. This group rating depends on the engine size and make of the car. i.e. some foreign cars are more expensive to insure.

The tables below show how many points are given with respect to group rating, type of cover, age of car, age and experience of driver and the use of the car.

*INEXPERIENCE—
means a driver aged 17/18 or aged 19+ with a provisional licence or who has passed the driving test less than one year ago.

GROUP RATING	I	II	III	IV	V	VI
POINTS	0	5	12	17	21	29

COVER	COMPREHENSIVE	THIRD PARTY FIRE & THEFT	THIRD PARTY ONLY
POINTS	23	5	1

AGE OF CAR	UP TO 3 YRS	4/5 YRS	6/7 YRS	8/9 YRS	10 YRS+
POINTS	14	13	12	11	9

AGE OF DRIVER	17/18	19	20	21	22	23	24	25/29	30/70
EXPERIENCE: PTS		24	17	14	12	9	6	4	2	0
INEXPERIENCE: PTS	24	24	21	19	16	13	11	9	7

USE OF CAR	SOCIAL DOMESTIC & PLEASURE	BUSINESS
POINTS	2	10

PERMITTED DRIVERS	ANY DRIVER	HUSBAND/WIFE	OWNER DRIVER
POINTS	6	5	4

40 Car Insurance

Example Jim Allen is an experienced driver aged thirty five. The car he owns is a 4 year old car in group III, he has a comprehensive insurance for himself and his wife and intends to use the car for social domestic and pleasure purposes. How many points would he be given?

Information	Points
Group III	12
Comprehensive	23
Age of car	13
Age of driver	0
Use of car	2
Permitted drivers	5
Total	55

Exercise 40.1

1. How many points would be given to an eighteen year old learner driver with a seven year old car in group I? Comprehensive insurance is required for any driver and the car is to be used for social domestic and pleasure purposes.

2. How many points would be given to a 25 year old experienced driver with a new car in group V? He intends to use the car for his business and requires comprehensive insurance for himself and his wife.

3. A forty year old experienced driver requires owner/driver third party fire and theft insurance. How many points would he be given if his 8 year old car is in group III and he intends to use the car for social domestic and pleasure purposes?

The tables below show the premiums required for certain points totals and details of no-claims discounts available.

POINTS	PREMIUM	POINTS	PREMIUM
16–18	70·40	49–51	169·20
19–21	75·60	52–54	184·60
22–24	81·00	55–57	200·40
25–27	87·20	58–60	229·60
28–30	91·60	61–63	272·40
31–33	97·00	64–66	320·80
34–36	102·20	67–69	397·00
37–39	107·60	70–72	469·20
40–42	123·00	73–75	555·60
43–45	138·60	76–78	623·20
46–48	153·80	79–81	748·60

YRS OF NO CLAIMS	PERCENTAGE DISCOUNT
1	20
2	30
3	40
4	50
5+	60

40 Car Insurance

Example John Evans is 24 years old, an experienced driver and requires comprehensive insurance for his 4 year old car which falls into group IV. John requires insurance only for himself and had 4 years of no claims. Find his net premium if he uses the car for social domestic pleasure purposes.

Information	Points
Group IV	17
Comprehensive	23
Age of car	13
Age of driver	4
Use of car	2
Permitted drivers	4
Total	63 points

From the table his gross premium is £272·40
No claims for 4 years = 50%
So his discount = 50% of £272·40 = £136·20
Net premium (what he actually pays) = £272·40 − £136·20 = £136·20

Exercise 40.2
Find the net premium due in the following cases.

1. Colin Smith, commercial traveller aged 45, requires comprehensive insurance for his new car in group III. Colin requires cover for any driver and has three years no claims.

2. Carol Hart is aged 20 with two years driving experience. Carol has a 5 year old car in group I and requires third party only with cover for any driver and uses the car for social, domestic and pleasure purposes. Carol has two years no claims.

3. Eric Holt, aged 29, is an experienced driver. Eric has comprehensive insurance on his 10 year old car in group V with cover for his wife to drive and has seven years no claims.

4. Sheila Brown, aged 31, has 12 years driving experience and requires comprehensive insurance on a two year old car in group VI. She wants owner/driver cover and itends to use the car for business. Sheila has 11 years no claims.

For discussion

1. Why is the age of a driver an important factor in determining insurance premiums?

2. What factors determine the group ratings given to certain cars?

3. What is the advantage of comprehensive insurance as opposed to third party and third party, fire and theft?

Projects

1. Find out the costs of insurance of cars belonging to parents, relatives, teachers etc.

2. Find out about motorbike insurance.

41 Car and Van Hire

There are usually two methods of payment involved in renting or hiring a car or van. Some firms offer a daily charge with unlimited mileage. Others make additional charges for excess mileage.

The table below shows charges for unlimited mileage.

GROUP AND MODEL	UNLIMITED MILEAGE—DAILY RATES				
	22 PLUS DAYS	15–21 DAYS	7–14 DAYS	3–6 DAYS	1–2 DAYS
A Mini	£7·75	£8·50	£9·00	£10·25	£10·75
B Fiesta/Metro	£10·75	£11·25	£11·75	£13·50	£14·25
C Sierra	£12·75	£13·25	£14·00	£16·00	£17·00
D Princess	£18·00	£18·75	£20·00	£23·00	£24·50
E Rover 2·3	£25·00	£27·50	£30·00	£36·00	£40·00
F Triumph TR7	£37·00	£39·50	£42·00	£48·00	£52·00
NOTE: VAT AT 15% ON ALL RENTALS					

Example How much would it cost to hire a car in group A for 4 days?
From the chart this rate is £10·25 per day.
Hire cost = £10·25 × 4 = £41·00
+VAT (15%) using calculator

Press

Display reads 47.15

Total hire cost = **£47·15**

Exercise 41.1

1. Find the cost of hiring the following:
   (a) A Sierra for 8 days.
   (b) A car in group D for 30 days.
   (c) A Rover 2·3 for 4 days.
   (d) A car in group F for 8 days.
   (e) A mini for 16 days.

2. Find the difference in costs between hiring:
   (a) A mini or a Fiesta for 6 days.
   (b) A Sierra or a Metro for 8 days.
   (c) A Rover 2·3 or a Triumph TR7 for 5 days.

3. Which is the most expensive and by how much:
   (a) Hiring a car in group B for 6 days or
   (b) Hiring a car in group A for 10 days?

41 Car and Van Hire

Van Hire

The table shows charges for van hire. This time only 50 miles per day on average are allowed in the cost. Distances more than 50 miles are charged at 6p or 7p per mile according to the type of van hired.

SHERPA VAN	FORD TRANSIT
Daily rates	Daily rates
£11·00 for 20–30 days	£17·00 for 20–30 days
£12·00 for 10–19 days	£18·00 for 10–19 days
£13·00 for 5–9 days	£20·00 for 5–9 days
£14·00 for 2–4 days	£22·00 for 2–4 days
£15·00 for 1 day	£23·00 for 1 day
Includes 50 free miles per day.	Includes 50 free miles per day.
Plus 6p per excess mile.	Plus 7p per excess mile.

NOTE: VALUE ADDED TAX (VAT) AT 15% ON ALL RENTALS

Example Find the cost of hiring a Transit for 6 days and travelling 700 miles.

		£
Hire cost	= £20·00 × 6	= 120·00
Excess mileage	= Actual − free	
	= 700 − (6 × 50)	
	= 700 − 300	
	= 400 miles	
Cost	= 400 × 7p	= 28·00
	Cost so far	148·00
	+VAT at 15%	
	(use calculator)	

Press

[1] [4] [8] [×] [1] [5] [%] [+] [=]   Display reads $\boxed{170.2}$

So, total cost = **£170·20**

Exercise 41.2

1. Find the total cost of hiring the following:

	Van	Days	Mileage
(a)	Sherpa	8	600
(b)	Transit	12	700
(c)	Sherpa	11	500
(d)	Transit	25	1000
(e)	Transit	3	1050
(f)	Sherpa	6	600
(g)	Transit	30	2500

2. A firm hire a Transit for 10 days and cover 850 miles. How much would they have saved by hiring a Sherpa for the same time and mileage?

3. Which is the most expensive and by how much?
   (a) Hiring a Sherpa for 9 days and driving 620 miles.
   OR
   (b) Hiring a transit for 7 days and driving 850 miles.

Answers

Chapter 1

Exercise 1.1
1. 428
2. 181
3. 224
4. 6290
5. 456885
6. 34
7. 340
8. 45
9. 1368
10. 3060

Chapter 2

Exercise 2.1
1. (a) 14000
   (b) 13800
   (c) 13770
2. (a) 43000
   (b) 16000
   (c) 20800
   (d) 23200

Exercise 2.2
1. 1600
2. 2
3. 6300
4. 40
5. 120
6. 50
7. 150000
8. £42·00
9. £99·00
10. 10

Exercise 2.3
1. £3·80
2. (a) 3 tins
   (b) £36·75
3. (a) 4
   (b) 8
4. (a) 10 boxes
   (b) £62·00
5. £2·30

Chapter 3

Exercise 3.1
1. £3·25
2. £3·75
3. £7·70
4. £4·25
5. £4·80
6. £4·80
7. £15·20
8. £7·80

Chapter 4

Exercise 4.1
1. 1255
2. 2025
3. 2325
4. 150

Exercise 4.2
1. (a) 570
   (b) 840
   (c) 100
   (d) 240
2. 395
3. 1065
4. (a) 15 minutes
   (b) 45 minutes

Chapter 5

Exercise 5.1
1. $\frac{3}{8}$
2. $\frac{1}{4}$
3. $\frac{5}{6}$
4. $\frac{3}{5}$
5. $\frac{4}{7}$
6. $\frac{3}{8}$
7. $\frac{5}{6}$
8. $\frac{5}{8}$
9. 11
10. 7

Exercise 5.2
1. $1\frac{1}{2}$
2. $1\frac{2}{5}$
3. $1\frac{1}{7}$
4. $1\frac{7}{8}$
5. $7\frac{1}{2}$
6. $3\frac{1}{7}$
7. $2\frac{1}{15}$
8. $5\frac{3}{10}$
9. $1\frac{17}{100}$
10. $18\frac{1}{3}$

Exercise 5.3
1. $\frac{3}{2}$
2. $\frac{9}{5}$
3. $\frac{11}{5}$
4. $\frac{19}{10}$
5. $\frac{15}{8}$
6. $\frac{11}{4}$
7. $\frac{7}{2}$
8. $\frac{14}{3}$
9. $\frac{50}{7}$
10. $\frac{50}{3}$

Exercise 5.4
1. $\frac{1}{2}$
2. $\frac{4}{5}$
3. $\frac{2}{3}$
4. $\frac{3}{4}$
5. $\frac{5}{12}$
6. $\frac{1}{4}$
7. $\frac{3}{4}$
8. $\frac{1}{3}$
9. $\frac{2}{5}$
10. $\frac{2}{3}$
11. $\frac{4}{5}$
12. $\frac{5}{12}$
13. $\frac{1}{2}$
14. $\frac{1}{3}$
15. $\frac{5}{8}$
16. $\frac{5}{3}$ or $1\frac{2}{3}$
17. $\frac{7}{4}$ or $1\frac{3}{4}$
18. 4
19. (a) $\frac{2}{5}$
    (b) $\frac{3}{5}$
20. 120

Chapter 6

Exercise 6.1
1. $\frac{2}{4}$
2. $\frac{4}{10}$
3. $\frac{2}{20}$
4. $\frac{70}{100}$
5. $\frac{4}{12}$
6. $\frac{6}{9}$
7. $\frac{18}{21}$
8. $\frac{6}{16}$
9. $\frac{15}{18}$
10. $\frac{28}{48}$
11. $\frac{90}{100}$
12. $\frac{27}{42}$
13. $\frac{24}{36}$
14. $\frac{45}{81}$
15. $\frac{84}{144}$
16. $\frac{143}{169}$
17. $\frac{16}{12}$
18. $\frac{68}{16}$
19. (a) 90
    (b) 70
    (c) 20 more
20. (a) 10 metres
    (b) 35 metres
    (c) 3 metres

Exercise 6.2
1. $\frac{2}{3}$
2. $\frac{1}{2}$
3. $\frac{1}{2}$
4. $\frac{1}{6}$
5. $\frac{7}{8}$
6. $\frac{7}{10}$
7. $\frac{1}{12}$
8. $1\frac{3}{4}$
9. $2\frac{11}{12}$
10. $\frac{1}{8}$
11. $\frac{5}{8}$
12. $\frac{1}{25}$
13. $\frac{11}{16}$
14. $\frac{5}{8}$
15. $\frac{1}{3}$
16. $\frac{7}{12}$
17. $5\frac{7}{8}$ minutes
18. $4\frac{3}{20}$ litres
19. (a) $15\frac{1}{2}$ minutes
    (b) 2 minutes
20. (a) $3\frac{1}{2}$
    (b) $3\frac{1}{8}$
    (c) $\frac{3}{8}$

Exercise 7.1
1. 2
2. $1\frac{3}{4}$
3. $1\frac{1}{5}$
4. 9
5. $\frac{3}{16}$
6. $\frac{1}{5}$
7. $\frac{13}{16}$
8. $\frac{3}{4}$
9. $\frac{3}{4}$
10. $\frac{4}{5}$
11. $\frac{1}{8}$
12. $\frac{8}{15}$
13. $1\frac{1}{5}$
14. $1\frac{1}{2}$
15. $2\frac{2}{9}$
16. $4\frac{2}{3}$
17. (a) 80 minutes
    (b) 90 seconds
18. (a) $2\frac{3}{4}$ Km
    (b) $8\frac{1}{4}$ Km
19. (a) $2\frac{5}{8}$ gallons
    (b) $7\frac{7}{8}$ gallons
    (c) $3\frac{1}{2}$ gallons
20. (a) 30 minutes
    (b) 50 minutes
    (c) $\frac{7}{15}$

Exercise 7.2
1. 20
2. 5
3. 2
4. $1\frac{1}{8}$
5. $2\frac{2}{5}$
6. $2\frac{1}{2}$
7. $\frac{5}{14}$
8. $1\frac{1}{14}$
9. $\frac{10}{11}$
10. $\frac{4}{5}$
11. $1\frac{1}{8}$

12. $\frac{51}{64}$
13. 2
14. $3\frac{3}{16}$
15. $\frac{1}{2}$
16. $2\frac{1}{2}$
17. (a) £1·50
    (b) 10
18. (a) 16
    (b) 28
19. (a) $\frac{1}{10}$ metre
    (b) 18
20. (a) 20
    (b) 180

Exercise 7.3
1. 630
2. 201
3. 1200
4. 462
5. 650
6. 330
7. 1525
8. 850
9. 441
10. 360

Chapter 8

Exercise 8.1
1. 1·3
2. 0·73
3. 2·6
4. 3·52
5. 1·17
6. 0·4
7. 7·45
8. 0·12
9. 0·99
10. 1·84

Exercise 8.2
1. £0·17
2. £1·30
3. £1·11
4. £4·01
5. £6·60
6. £5·10
7. £0·53
8. £0·03
9. £6·65
10. £1·02

Exercise 8.3
1. 4·6
2. 8·2
3. 1·2
4. 1·16
5. 16·58
6. 10·61
7. 0·85
8. 2·9
9. 6·69
10. 9·39
11. 2·81
12. 1·92
13. 3·59
14. 3·16
15. 2·12

16. 4·28
17. 4·36
18. 1·75
19. (a) £3·22
    (b) £1·78
20. 2·91

Exercise 8.4
1. 19
2. 2·5
3. 6·2
4. 99·7
5. 837·5
6. 607·5
7. £16·50
8. 446·5
9. 2006
10. £12·00

Exercise 8.5
1. 17·1
2. 26·4
3. 0·84
4. 287·2
5. 2·04
6. 1·32
7. 3·28
8. 37·32
9. (a) £8·54
    (b) £13·42
10. (a) £60·00
    (b) 12 miles

Exercise 8.6
1. 0·27
2. 1·24
3. 0·56
4. 1·15
5. 0·19
6. 0·06
7. £0·66
8. 0·016
9. 0·62
10. £0·07

Exercise 8.7
1. 0·24
2. 3·5
3. 1·2
4. 91
5. 6
6. 2·75
7. 0·41
8. 5
9. £5·58
10. 41

Chapter 9

Exercise 9.1
1. 0·5
2. 0·75
3. 0·4
4. 0·125
5. 0·875
6. 0·8
7. 0·05
8. 0·55
9. 0·04
10. 0·16

Exercise 9.2
1. $0·\dot{6}$
2. 0·625
3. 0·95
4. $0·8\dot{3}$
5. $0·\dot{1}$
6. 0·3125
7. $0·\dot{5}$
8. $0·\dot{7}$
9. $0·91\dot{6}$
10. 0·21875

Exercise 9.3
1. 0·17
2. 0·69
3. 0·71
4. 0·14
5. 0·29
6. 0·63
7. 0·95
8. 0·29
9. 0·92
10. 0·89

Chapter 10

Exercise 10.1
1. $\frac{1}{4}$
2. $\frac{1}{2}$
3. $\frac{13}{20}$
4. $\frac{4}{5}$
5. $\frac{17}{20}$
6. $\frac{4}{25}$
7. $\frac{27}{100}$
8. $\frac{13}{5}$
9. $\frac{9}{25}$
10. $\frac{49}{100}$

Exercise 10.2
1. 75%
2. 90%
3. 5%
4. $33\frac{1}{3}$%
5. 40%
6. 55%
7. 20%
8. (a) 60%
    (b) 40%
9. (a) 45%
    (b) 55%
10. (a) 20%
    (b) 5%
    (c) 30%

Exercise 10.3
1. 36
2. 12
3. 110
4. £140·00
5. 8
6. £100·00
7. 56 Seconds
8. £80·00
9. (a) £8·00
    (b) £88·00
10. (a) £36·00
    (b) £276·00

Exercise 10.4
1. £9·35
2. 14·04

3. £96·72
4. 2·1
5. £14·08
6. 68·59
7. £2·34
8. 58·32
9. 13·56
10. 82·875

Exercise 10.5
1. (a) £18·75
    (b) £4557·00
2. (a) £492·80
    (b) £468·75
3. (a) £5306·40
    (b) £6700·00
4. (a) £245·00
    (b) £136·50
5. (a) £97·98
    (b) £140·58
6. £1147·00
7. (a) £1·67
    (b) £1·04
8. £111·$82\frac{1}{2}$
9. 1325000
10. (a) 2190000
    (b) 2398050

Chapter 11

Exercise 11.1
1.
2. (a) 2·47 m
    (b) 0·75 m
    (c) 0·09 m
3. (a) 50 cm
    (b) 75 cm
    (c) 40 cm
4. 4·977 m
5. (a) $\frac{1}{2}$
    (b) $\frac{3}{10}$
6. 67 kg
7. (a) £3·75
    (b) £0·50
8. (a) 250 ml
    (b) 100 ml
9. (a) $\frac{1}{5}$
    (b) $\frac{1}{2}$
10. (a) 200
    (b) 70
11. (a) 30
    (b) 100
12. (a) 10 kg
    (b) 30
13. (a) 144 ml
    (b) 100 ml
14. (a) 200 g
    (b) 350 g
    (c) 15 g
    (d) $1\frac{3}{4}$ litres

Exercise 11.2
1. (a) 4 in
    (b) 18 in
    (c) 49 in
2. (a) 12 oz
    (b) 10 oz
    (c) 56 oz
3. (a) 6 pints
    (b) 11 pints
    (c) 47 pints

4. (a) 880 yds
   (b) 220 yds
   (c) 176 yds
5. $9\frac{1}{2}$ in
6. 21 gallons
7. 34 oz
8. (a) £3·57$\frac{1}{2}$
   (b) £3·96
9. (a) 52 lb
   (b) 16 stones
10. (a) 360 miles
    (b) $9\frac{1}{2}$ gallons

Exercise 11.3
1. (a) $12\frac{1}{2}$ cm
   (b) 24 km
   (c) 9 kg
2. (a) 4·8 cm
   (b) 40 miles
   (c) 110 lb
3. 80 cm
4. 42 in
5. 6 ft
6. (a) 27°C
   (b) 50°F
   (c) 4°C
7. (a) 98°F
   (b) 85°F
   (c) 13°C
8. (a) 1·54 gallons
   (b) 5·06 gallons
   (c) 18·26 gallons
9. (a) £3·90
   (b) £8·86$\frac{1}{2}$
   (c) £26·59$\frac{1}{2}$
10. (a) 32 litres
    (b) £12·16

Chapter 12

Exercise 12.1
1. (a) 50 minutes
   (b) 1 hour 35 minutes
   (c) 35 minutes
   (d) 50 minutes
2. (a) 06·52
   (b) 15·46
   (c) 19·59
   (d) 23·45
   (e) 02·40
3. (a) 7·45 a.m.
   (b) 12·15 a.m.
   (c) 11·45 p.m.
   (d) 8·05 p.m.
   (e) 9·17 a.m.
4. (a) 9 minutes fast
   (b) 4 minutes slow
   (c) 3 minutes slow
   (d) 8 minutes fast
   (e) 13 minutes slow
5. (a) 10
   (b) 19
   (c) 15
   (d) 21
6. (a) 211
   (b) 76
   (c) 158
   (d) 90
7. (a) 13·20
   (b) 10·17
8. (a) 07·45
   (b) 16·45
   (c) 08·20
   (d) 10·20
9. (a) 16·15
   (b) 00·15
   (c) 08·15
   (d) 05·15
10. (a) 10·45
    (b) 15·45
    (c) 07·00

Exercise 12.2
1. 42 minutes
2. (a) 09·33
   (b) 10·15
3. 36 minutes
4. 12 minutes
5. (a) 14·40
   (b) 1 hour 10 minutes
6. 20 minutes
7. 12·57
8. 2 hours 15 minutes
9. (a) 19·14
   (b) 19·20
10. (a) 33 minutes
    (b) 2 hours 27 minutes

Chapter 13

Exercise 13.1
1. 200 g
2. £1000 and £1400
3. 25 kg and 30 kg
4. 150 females
5. £3500, £7000, £10500
6. 20 years
7. £35
8. 360 g
9. (a) £18 and £24
   (b) £72
10. 10 kg meat, 20 kg vegetables

Exercise 13.2
1. £15
2. £819
3. 245 miles
4. (a) £25
   (b) 16 hours
5. (a) 2189 pesetas
   (b) £20
6. £238·15
7. (a) £9·36
   (b) 6·5 kg
8. (a) $13·875
   (b) £7·50
9. (a) 455·1
   (b) £81·30
10. (a) 173·7
    (b) 81p

Chapter 14

Exercise 14.1
1. (a) 350
   (b) 550
   (c) 1970 to 1980
2. (a) 30000 litres
   (b) 60000 litres
   (c) 175000 litres
   (d) 35000 litres
3. (a) 35
   (b) 85
   (c) 60
   (d) 1977 to 1979
4. (a) 55
   (b) 85
   (c) 175
   (d) 1977
5. (a) Thursday
   (b) £3150
   (c) £525
6. (a) 63°
   (b) 90°
   (c) 48
   (d) 16
7. (a) 3600
   (b) 21600
   (c) 7200
   (d) 360
8. (a) 15 miles
   (b) 48 km
   (c) 60 miles
   (d) 72 km
9. (a) 44 lbs
   (b) 18 kg
   (c) 88 lb
   (d) 46 kg
10. (a) 38°
    (b) 39°
    (c) Monday p.m.
    (d) Wednesday a.m. to Thursday a.m.

Chapter 15

Exercise 15.1
1. (a) 21 cm^2
   (b) 18·4 cm
2. (a) 3·5 m
   (b) 15 m
3. (a) 4 cm
   (b) 20 cm^2
4. (a) 7·6 m^2
   (b) 23·2 m^2
5. (a) 8 mm
   (b) 8 m
6. (a) 251·1 cm^2
   (b) 19·375 m^2
   (c) 102·5 cm^2
7. (a) 10·54 m
   (b) 17·36 cm
   (c) 73·78 cm
8. (a) 22 m^2
   (b) 1400 cm^2
   (c) 518·4 cm^2
   (d) 2295 mm^2
9. (a) 334·8 cm^2
   (b) 1739·1 mm^2
10. (a) 168
    (b) 14
    (c) £52·50

Chapter 16

Exercise 16.1
1. (a) 32000 mm^3
   (b) 15 m^3
   (c) 3750 cm^3
2. 900 cm^3
3. (a) 1008 m^3
   (b) 30000 mm^3
   (c) 24 cm^3
4. 99·2 m^3
5. (a) 1·55 m^2
   (b) 38750 mm^3
6. 8 cm
7. 20 boxes
8. 1183600 mm^3
9. (a) 136000 cm^3
   (b) 118000 mm^3
10. 3·5 m^3

Chapter 17

Exercise 17.1
2. (a) 8·75 cm^2
   (b) 7 cm^2
   (c) 5 m^2
   (d) 10 m^2
3. (a) 14 cm^2
   (b) 50·75 m^2
4. £1080
6. (a) 35 km
   (b) 15 km
   (c) 35 km
   (d) 50 km
7. (a) 70 km
   (b) 350 km
   (c) 25 litres
   (d) £9·25
8. (a) 370 miles
   (b) 850 miles
   (c) 480 miles
9. 1240 miles
10. (a) 385 mph
    (b) 520 mph
    (c) 200 mph

Chapter 18

Exercise 18.1
1. (a) 624
   (b) 300
   (c) 924
2. (a) £13·50
   (b) £4·00
   (c) £18·00
   (d) £3968
3. (a) 708
   (b) 216
   (c) £3
4. (a) 149
   (b) £3500
5. (a) 780
   (b) £4200
   (c) 180

Chapter 19

Exercise 19.1
1. (a) 24 m
   (b) 72 cm^2
   (c) 288 m^2
2. (a) 48 m^2
   (b) 65 m^2
   (c) 48 m^2

      (d) 27 m²
      (e) 40 m²
3. (a) 11·25 m²
   (b) 246
   (c) £513
4. (a) 5
   (b) £58·50
5. (a) 180
   (b) £66·60

Chapter 20

Exercise 20.1
1. (a) 5022 cm²
   (b) 10044 cm²
   (c) 2034 cm²
2. (a) £72·80
   (b) £63·35
   (c) £53·55
3. (a) 48
   (b) 108
   (c) 228
4. (a) 10%
   (b) £2640
5. (a) £3300
   (b) 6%

Chapter 21

Exercise 21.1
1. (a) 12:1
   (b) 20:1
   (c) 5:3
2. (a) 150 kg
   (b) 12½ kg
   (c) 7½ kg
4. (a) 72 kg
   (b) 48 kg
   (c) 60 kg
   (d) 24 kg
   (e) 100 dozen
   (d) 10:3
5. (a) Mark 4
   (b) 325°F
   (c) 425°C
   (d) Mark 6

Chapter 22

Exercise 22.1
1. 45°
2. 30°
3. 75°
4. 120°
5. 150°

Exercise 22.2
1. (a) A
3. (a) A

Exercise 22.3
1. (a) 060°
   (b) 120°
   (c) 095°
   (d) 210°
2. (a) 050°
   (b) 140°
   (c) 060°
   (d) 120°

Chapter 23

Exercise 23.1
1. (a) £84
   (b) £56
2. (a) £25·20
   (b) £15·20
3. £109·20
4. (a) 6 hours
   (b) 3 hours
   (c) £154
5. (a) Clerical
   (b) £29·63
   (c) £63·47
8. £141·50
9. (a) £350
   (b) £240
   (c) 630
   (d) £490
10. (a) £5875
    (b) £7800
    (c) £9500

Chapter 24

Exercise 24.1
1. (a) £312·31
   (b) £290·11
   (c) £129·71
   (d) £57·15
2. £5·35
3. £590·15

Chapter 25

Exercise 25.1
1. £426·60
2. £6937·50
3. £3143·40
4. £1054·88
5. £1713·90
6. £335·09
7. £491·11
8. £2671·38
9. £159·38
10. £1280·32

Exercise 25.2
1. £1733·44, £233·44
2. £3300·09, £900·09
3. £32913·32, £7913·32
4. £810·93, £270·93
5. £210·54, £90·54
6. £64218·28, £14218·28
7. £1638·91, £383·91
8. £1548816·80, £548816·18
9. £5600·55
10. £255·36

Chapter 26

Exercise 26.1
1. (a) £32·20
   (b) £52·80
2. (a) £303·20
   (b) £48·20

Exercise 26.2
1. (a) £70
   (b) £65
   (c) £39
   (d) £650
   (e) £595

Exercise 26.3
1. (a) £55·24
   (b) £76·20
   (c) £110·24
   (d) £121·44
2. (a) £92·00
   (b) £812·00
3. (a) £300
   (b) £2700
   (c) £99·36
4. (a) £350
   (b) £3150
   (c) £160·02
   (d) £115·92
5. £44·16

Chapter 27

Exercise 27.1
1. (a) £20·47
   (b) £25·76
   (c) £16·68
2. (a) £2·94
   (b) £22·54
4. (a) £0·45
   (b) £1·35
   (c) £4·50
   (d) £1·56
   (e) £0·85
5. (a) £41·17
   (b) £20·64
   (c) £16·96
   (d) £144·27
6. (a) £33
   (b) £8·25
   (c) £28·50
   (d) £1·69
7. (a) B by 75p
   (b) Y by £5·75
8. (a) 99½p
   (b) £10·94½
9. £72·10½
10. £164·45

Chapter 28

Exercise 28.1
1. (a) £19825
   (b) £2600 profit
   (c) £22425
   (d) £6150
2. (a) £91370
   (b) £1620 loss
   (c) £91370
   (d) £3790

Chapter 29

Exercise 29.1
1. (a) 4th April–13th May
   (b) 16th July–14th August
2. (a) £268·80
   (b) £350·80
3. (a) £336
   (b) £11·20
4. £174
5. £354
6. (a) 18·00
   (b) 20·10
   (c) 12·35
   (d) 02·00
7. 15·25
8. (a) 02·15
   (b) 03·00
   (c) 03·30
9. (a) October
   (b) 63°
   (c) 49°
10. (a) 5·8 hours
    (b) 2·9 hours
    (d) October, March and April
    (c) November, December and January

Chapter 30

Exercise 30.1
1. (a) 178·5
   (b) 178·8
   (c) 53000
   (d) 754·6
2. (a) £71·71
   (b) £604·03
   (c) £2512·56
   (d) £471·70
3. (a) £1·26
   (b) £1·76
   (c) £1·63
   (d) £1·38
4. (a) 142 francs
   (b) £13·17
5. (a) £48·18
   (b) 5926 drachmas
   (c) 3493 Belgian francs
6. (a) 8
   (b) 56
   (c) 134
   (d) 18
7. (a) £18·35
   (b) £44·03
   (c) £77·05
8. (a) Austria by £5·88
   (b) Germany by £1·40

Chapter 31

Exercise 31.1
1. (a) £7650
   (b) £8390
2. (a) £24250
   (b) £26250
3. (a) £5190
   (b) £20760
   (c) Yes

4. (a) £3100
   (b) £27900
   (c) No
5. (a) £13650
   (b) £31850
   (c) Yes

Exercise 31.2
1. (a) £154·80
   (b) £166·32
2. (a) £53460
   (b) £31104
3. (a) £24300
   (b) £4140
4. (a) £34000
   (b) £275·40

Chapter 32

Exercise 32.1
1. £51·50
2. £67·00
3. £152·00
4. £124·00
5. £143·00
6. £128·00
7. £435·00
8. £624·00

Chapter 33

Exercise 33.1
1. 1° = £50000
2. (a) 45°
   (b) 60°
   (c) 170°
3. (a) £8·5 million
   (b) £2·25 million
   (c) £2·5 million
4. (a) 12·5%
   (b) 5%

Exercise 33.2
1. 125p in the £
2. 135p in the £
3. 130p in the £
4. £28 million
5. £16·2 million

Exercise 33.3
1. £34·20
2. £409·20
3. £200
4. 160p in the £
5. £47·50

Chapter 34

Exercise 34.1
1. (a) 1224
   (b) £65·49
2. (a) 11785
   (b) 12441
   (c) 656
   (d) £38·06
3. (a) £84·23
   (b) £123·35
   (c) £98·00
4. £218·49
5. (a) £2·37
   (b) £17·63
   (c) £12·56

6. £62·79
7. (a) £0·58
   (b) £28·26
8. £213·49

Chapter 35

Exercise 35.1
1. (a) 5276
   (b) 9327
2. (a) 435
   (b) 366
   (c) 586
3. (a) 247
   (b) 255·398
   (c) £86·78
4. (a) 204
   (b) 211
   (c) £73·36
5. (a) £109·96
   (b) £71·22

Chapter 36

Exercise 36.1
1. (a) Cheap
   (b) Standard
   (c) Cheap
   (d) Standard
   (e) Cheap
2. (a) 10p
   (b) 75p
   (c) 35p
   (d) 40p
3. (a) 55p
   (b) 50p
4. (a) 20p
   (b) 30p
5. (a) Sutton & Penton
   (b) Newtown & Ashbridge
6. (a) 10p
   (b) £1·25
   (c) 75p
7. (a) £2·50
   (b) 35p
   (c) 10p
8. (a) 35p
   (b) 64p
   (c) 88p
   (d) £2·89
   (e) £2·79
   (f) £2·09

Chapter 37

Exercise 37.1
1. (a) 8 m
   (b) 88 m^2
2. (a) $17\frac{1}{2}$ m^2
   (b) $26\frac{1}{4}$ m^2
3. (a) $17\frac{1}{2}$ m^2
   (b) 2 litres
   (c) 34
   (d) 9 rolls
   (e) £39·42
4. (a) 3 litres
   (b) 44
   (c) 11
   (d) £57·03

5. (a) 55 m^2
   (b) £533·50
   (c) 220
   (d) 22
   (e) Carpet and underlay by £71·50

Chapter 38

Exercise 38.1
1. (a) £4906·25
   (b) £697·02
   (c) £5344·52
2. (a) £680·02
   (b) £729·14
3. £375·04

Exercise 38.2
1. £5351·91
2. £6535·04

Exercise 38.3
1. (a) £640
   (b) £3964·60
   (c) £764·60
2. (a) £2150
   (b) £3672·72
   (c) £472·72
3. (a) £2475
   (b) £1755
4. (a) £2520
   (b) £1920
   (c) £3792·92
5. (a) £1086·75
   (b) £1207·50

Chapter 39

Exercise 39.1
1. (a) £3767·86
   (b) £3120·16
2. (a) £2149·44
   (b) £3132·04
3. £1609·98
4. 5 years
5. 3 years

Exercise 39.2
1. (a) 10
   (b) 15
   (c) 5
   (d) 20
2. (a) 8
   (b) 10
   (c) 90 km/hour
   (d) 60 litres
3. (a) 8 litres/100 km
   (b) £3
   (c) £156
4. (a) £1653·75
   (b) £2200·00
   (c) £42·31

Chapter 40

Exercise 40.1
1. 67
2. 75
3. 34

Exercise 40.2
1. £192·48
2. £71·54
3. £108·96
4. £299·44

Chapter 41

Exercise 41.1
1. (a) £110·40
   (b) £621
   (c) £165·60
   (d) £386·40
   (e) £156.40
2. (a) £22·42
   (b) £20·70
   (c) £69·00
3. Hiring in A by £10·35

Exercise 41.2
1. (a) £133·40
   (b) £256·45
   (c) £151·80
   (d) £488·75
   (e) £148·35
   (f) £110·40
   (g) £667
2. £69·92
3. Hiring a transit by £54·97